International Conference on Advances in Materials Science, Communication and Microelectronics (ICAMCM 2021)

IOP Conference Series: Materials Science and Engineering
Volume 1119

Jaipur, India
19 - 20 February 2021

Editors:

V.K. Chandna
Sandeep Vyas
Girraj Sharma
Ashish Kumar

ISBN: 978-1-7138-3405-2
ISSN: 1757-8981

Printed from e-media with permission by:

Curran Associates, Inc.
57 Morehouse Lane
Red Hook, NY 12571

Some format issues inherent in the e-media version may also appear in this print version.

Institute of Physics
Dirac House, Temple Back
Bristol BS1 6BE UK

Phone: 44 1 17 929 7481
Fax: 44 1 17 920 0979

techtracking@iop.org

Additional copies of this publication are available from:

Curran Associates, Inc.
57 Morehouse Lane
Red Hook, NY 12571 USA
Phone: 845-758-0400
Fax: 845-758-2633
Email: curran@proceedings.com
Web: www.proceedings.com

TABLE OF CONTENTS

IOP Conf. Series: Materials Science and Engineering 1119 (2020) 011001 doi:10.1088/1757-899X/1119/1/011001

JAIPUR ENGINEERING COLLEGE AND RESEARCH CENTRE

International Conference on Advances in Material Science, Communication and Microelectronics (ICAMCM-2021)
(February 19-20, 2021)

In Association with

IOP: Materials Science and Engineering (Scopus Indexed)

Technically Sponsored by

OSA Student Chapter-JECRC
&
IETE Rajasthan Centre

Organized By

Jaipur Engineering College and Research Centre
Department of Electronics and Communication Engineering
Shri Ram ki Nangal, via Sitapura RIICO, Tonk Road,
Jaipur, Rajasthan 302022

IOP Conf. Series: Materials Science and Engineering 1119 (2020) 011001 doi:10.1088/1757-899X/1119/1/011001

Editors of ICAMCM-2021 for IOP: Materials Science and Engineering Proceeding are:

- Prof. V. K. Chandna
  Principal, JECRC,
  General Chair ICAMCM-2021

- Dr. Sandeep Vyas
  HOD ECE, JECRC,
  Conference Chair, ICAMCM-2021

- Dr. Girraj Sharma
  Associate Professor, ECE Department, JECRC
  Organizing Chair, JECRC, Jaipur

- Dr. Ashish Kumar
  Assistant Professor, ECE Department, JECRC
  Organizing Chair, JECRC, Jaipur

About the Conference: The International Conference on Advances in Materials Science, Communication and Microelectronics (ICAMCM-2021) is to be organized with the perception to address the various issues of the society in the fields of material science, communication engineering and electronic technology. In this conference, it will be expected that researchers will bring new prospects for collaboration across intra and inter disciplines to facilitated novel concepts. This premier conference will be organized by the Department of Electronics & Communication Engineering during 19-20 February 2021 at Jaipur Engineering College and Research Centre, Jaipur. The conference is extended over different areas of engineering & technology, mainly focusing on the field as mentioned above. The idea of organizing the conference is to bring together the researchers, scientists, engineers, academicians, and industrialists in areas of Engineering and Technology, on a common international forum for the dissemination of original research, new ideas, outcomes, and challenges. It will provide an opportunity for the global participants to share their ideas and experience in person with their peers expected to join from different parts of the world. In addition, this gathering will help the delegates to establish research or business relations as well as to find international linkage for future collaborations in their career path.

The conference would offer a large number of invited lectures from renowned speakers from all over the world. The theme of this conference will motivate the researchers to adopt the outcome for implementation. This Conference is technically sponsored by the Optical Society of America (OSA) Student Chapter-JECRC and IETE Rajasthan Centre.

About Institute: A journey of Two decades for JECRC, having more than 4000 students on campus under 6 UG programs, has earned laurels to their students, faculty members and the institute in many ways. More than 10000 alumni's spread over the globe has climbed the

IOP Conf. Series: Materials Science and Engineering 1119 (2020) 011001 doi:10.1088/1757-899X/1119/1/011001

ladder to leadership positions and providing mentorship to their juniors by way of skill development, incubation, startup, research, and angel funding. Faith by government agencies for providing a grant of more than 2 crores for setting up a centre of excellence, state of art facilities for startup & incubation, and providing a platform to the students to develop their technical and managerial skills that are helping students to get placement in a reputed organization. Contribution towards International publications, technical activities, co-curricular activities by faculty members, students, and delivery of outcome-based education is recognized by the National Board of Accreditation and AICTE. Socially rich atmosphere at the campus enabling fourfold grooming of students that is recognized at National and International level and enabling students to work as interns with personalities recognized in their field of expertise. JECRC has become synonymous to placements and JECRCians have made their presence felt at every reputed company/government organization.

About ECE department: The Electronics and Communication Engineering (ECE) Department has a rich tradition in research and teaching. The ECE department is NBA accredited. The research interests of the faculty members of the department encompass the wide area of applied and fundamental aspects of Electronics and Communication Engineering. Collaboration between the research groups of the department and with other departments is also a practice. The broad areas of research and academic activities in the department are:

- Communication Systems
- Microelectronics and VLSI
- RF and Microwave Engineering

Proposed Tracks:

1. Material Science, VLSI, MEMS and nanotechnology

- Advances in Nano Materials Science
- Materials Synthesis, Processing and application
- Materials Science-Fundamentals
- Emerging Areas of Materials Science
- Process development of MEMS/ NEMS devices
- MEMS/NEMS sensors and actuators
- Organic and flexible electronic
- Bio/Nano Sensing
- Biological Interface cells at the Nanoscale
- Frontiers in nanobiotechnology

2.Digital signal, Image and video processing

- Processing and Analysis of Optical and Microwave Sensing
- Images
- Satellite data acquisition technology
- Audio and Acoustic Signal Processing
- Speech Processing
- Quantum Signal Processing
- Remote Sensing and Signal Processing
- Compressive Sensing, Sampling, and Dictionary Learning
- Sensor Array & Multichannel Signal Processing
- Image, Video and Multidimensional
- Signal Processing for Cyber Security

3.RF, wireless and Optical Fiber Communication
- 4G&5G
- 802.16 & WiMAX
- Antenna Design (Lower band / Higher Band)
- Antenna Systems for Futuristic Applications
- Cognitive radio
- OFDM and MIMO techniques
- Telecommunications & Radio-frequency engineering
- Optical Communication
- Photonic devices and systems

4.Machine learning and IOT based engineering solutions
- Applied computing
- Articial intelligence & Machine learning
- Computer architecture & security
- Cloud Computing & Cluster Computing
- Internet of Things (IOT)
- Software Engineering
- Cryptography & Cybernetics

5.Real time-based embedded systems
- VLSI Technology & Architecture
- Bioinformatics & Applications
- Digital signal processors and microcontrollers
- Digital computers
- Instrumentation, and microelectronics
- Sustainability challenges in Emerging Technologies

Organizing Committee Members:

Patrons

- Shri O.P. Agrawal, Chairman, JECRC Foundation
- Shri M. L. Sharma, Vice Chairman, JECRC Foundation
- Shri Amit Agrawal, Vice-chairperson, JECRC Foundation
- Shri Arpit Agrawal, Vice-chairperson, JECRC Foundation

General Chair

- Prof. V. K. Chandna, Principal, JECRC

Conference Chair

- Dr. Sandeep Vyas, HOD, Department of ECE, JECRC

Organizing Chair

- Dr. Girraj Sharma Associate Professor, JECRC
- Dr. Ashish Kumar, Assistant Professor, JECRC

Organizing Secretary

- Dr. Parul Tyagi, Associate Professor, JECRC
- Dr. Jaiverdhan, Assistant Professor, JECRC
- Mr. Bhupesh Kumawat, Assistant Professor, JECRC

IOP Conf. Series: Materials Science and Engineering 1119 (2020) 011001 doi:10.1088/1757-899X/1119/1/011001

Advisory Board (International)

- Prof. Bimal K Bose, Uni. of Tennessee, USA
- Prof. Muhammad H Rashid, Uni. of West Florida, USA
- Dr. Chi Hieu Le, Uni. of Greenwich, United Kingdom
- Dr. Raphael Guerrero, Ateneo de Manila Universty, Philippines
- Dr. Umesh Chand, National University of Singapore, Singapore
- Dr. Sha Shiong Ng, Malaysia
- Dr. Tawfik Ismail, Cairo University, Egypt
- Prof. Mourad Zghal, Tunisia
- Prof. Anand Nayyar, Duy Tan University, Vietnam

Advisory Board (National)

- Prof. (HAG) R. P. Yadav, MNIT Jaipur
- Prof. M. M. Sharma, MNIT Jaipur
- Prof. Vijay Janyani, MNIT Jaipur
- Prof. Ghanshyam Singh, MNIT Jaipur
- Dr. Mahanth Prasad, Principal Scientist, CSIR-CEERI, Pilani
- Dr. Ritu Sharma, MNIT Jaipur
- Dr. Soumava Mukherjee, IIT Jodhpur, India
- Shri O.P. Jain, JECRC Jaipur
- Shri P. K. Tiwari, JECRC Jaipur
- Prof. S.N. Gupta, JECRC Jaipur
- Mr. Manish Jain, JECRC Jaipur
- Prof. Mukt Bihari, JECRC Jaipur
- Prof. Utpal Chakravatky, JECRC Jaipur
- Technical Committee
- Dr. Kuldeep Singh, MNIT Jaipur
- Dr. Rajendra Mithrwal, MNIT Jaipur
- Dr. Sarthak Singhal, MNIT Jaipur
- Dr. Ravi Maddila, MNIT Jaipur
- Dr. Deepak Sharma, CSIR CSIO Delhi
- Dr. Anup Kumar Sharma, NIT Silchar
- Dr. Nikhildeep Gupta, NIT Nagpur
- Dr. Arathy Varghese, IIT Bombay
- Dr. Satyanarayan Patel, IIT Indore
- Dr. Vipin Pal, NIT Meghalaya
- Dr. Sankey Saxena, LPU Jalandhar
- Dr. Amit Kumar Garg, IIIT Kota
- Dr. Vinay Kanungo, SKIT Jaipur
- Dr. Ashok Kumar Kajla, PCE, Jaipur
- Dr. Shashikant Sharma, IIIT Ranchi
- Dr. Aditi, CSIR CEERI Pilani
- Dr. Yogita, NIT Meghalaya
- Dr. Aman Kumar, NIT Hamirpur
- Dr. L D Sharma, VIT Andrapradesh
- Mr. Vipin Parkash Yadav, UD, RTU KOTA
- Dr. Kapil Saraswat, Central University, Rajasthan

IOP Conf. Series: Materials Science and Engineering 1119 (2020) 011001 doi:10.1088/1757-899X/1119/1/011001

- Dr. Anukul Pandey, DTU Delhi
- Dr. Mahendra Yadav, JP University
- Dr. Arun Kishor Johar, MNIT Jaipur
- Mr. Gaurav Sharma, MNIT Jaipur
- Mr. T. Bharat Kumar, MNIT Jaipur
- Mr. Vivek Upadhayay, ACEIT, Jaipur
- Mr. Surendra Saini, JNU Jaipur
- Dr. S. K. Singh, JECRC Jaipur
- Dr. Vinita Mathur, JECRC Jaipur
- Dr. Shruti Kalra, JECRC Jaipur

Organizing Committee

- Dr. S.S. Manakatala
- Mr. Anil Jain
- Mr. Lokesh Kr. Sharma
- Ms. Ritu Vyas
- Mr. Naresh Kumar
- Mr. Raj Kumar Jain
- Mr. Pravin Kr. Sharma
- Mr. Devendra Sharma
- Mr. Jitendra Sharma
- Mr. Sandeep Kr. Dotya
- Mr. Deepak Shankhala
- Mr. Honey Agarwal
- Mr. Rakesh Kardam
- Mr. Mangi Lal
- Ms. Deepmala Kulshrestha
- Ms. Yogita
- Mr. Mohit Rajput
- Ms. Mamta Rani
- Ms. Anju Rajpoot
- Ms. Tripti Dua

Publication Committee

- Dr. Neha Singh, JECRC Jaipur
- Ms. Ritambhara, JECRC Jaipur
- Mr. Devesh Gupta, JECRC Jaipur
- Ms. Nishi Atrary, JECRC Jaipur

Publicity Committee

- Mr. Ashish Sharma, JECRC Jaipur
- Mr. Ankur Gangwar, JECRC Jaipur
- Mr. Vikash Sharma, JECTRC Jaipur

Industry interaction sponsorship Committee

- Mr. Ashutosh Sharma

Web Team

- Mr. Naveen Kumar Kedia, JECRC Jaipur
- Ms. Avni Sharma, JECRC Jaipur

Hospitality Committee

- Mr. Ashish Kulshrestha, JECRC Jaipur
- Mr. J P Mishra, JECRC Jaipur

IOP Conf. Series: Materials Science and Engineering 1119 (2020) 011002 doi:10.1088/1757-899X/1119/1/011002

Peer review declaration

All papers published in this volume of *IOP Conference Series: Materials Science and Engineering* have been peer reviewed through processes administered by the Editors. Reviews were conducted by expert referees to the professional and scientific standards expected of a proceedings journal published by IOP Publishing.

- **Type of peer review: Single blind Review**

The detailed review process is as follows

1. Authors will submit their original and unpublished work (related to the theme of the conference). They clearly need to identify the problem, their contribution(s), and justification with respect to the state-of-the-art work.
2. Authors need to submit their papers through Easy Chair link.
3. The PC chairs will initially do the screening on the basis of scope, technical content, similarity and formatting of the submitted papers. The suitable papers will be sent for single-blind review.
   a. PC chairs will allow scientific committee members to choose the research topics they are willing to review. Based on this and their expertise, PC chairs will assign papers for a review manually.
   b. Scientific committee members can invite sub-reviewers and can hand over some papers if the papers are not within their field of expertise, and the whole process is available at the Easychair (i.e., taking help from sub reviewers).
   c. Scientific committee members are, however, still responsible for the quality and timeliness of the reviews allocated originally to them. Accordingly, the paper/review management system supports the allocation of reviews, at the initiation of PC members.
4. For evaluating the papers, the Easy Chair scores mechanism will be as followed:
   i. Strong accept (3), Accept (2) and Weak accept (1)
   ii. Borderline (0)
   iii. Strong reject (-3), Reject (-2) and Weak reject (-1)
5. In all cases, PC chairs will work closely. Based on the review reports and the scores, PC chairs send a notification to the authors via along with the review reports/scores.
   a. Notification is based on an average score provided by the assigned reviewers.
   b. Note that notification, in the first review, will be straight: accept and reject.
   c. The papers with an average score less than 0 will be rejected. For accepted papers, general chairs will provide a month for an additional revision, if any. This will mostly useful for those authors who have received major comments.
6. After having revised versions, we will have exactly same set of experts for review, and will add one more in case there exists contradictory review reports/scores. An example of contradictory review means giving +3 (strong accept) and -3 (strong reject) scores from two different for the same paper. At this point, PC chairs will help invite an additional reviewer.
7. After having all review reports/scores, PC chairs will make a final decision.

IOP Conf. Series: Materials Science and Engineering 1119 (2020) 011002 doi:10.1088/1757-899X/1119/1/011002

- **Conference submission management system: Easy-Chair**
- **Number of submissions received: 75**
- **Number of submissions sent for review: 52**
- **Number of submissions accepted:23**
- **Acceptance Rate (Number of Submissions Accepted / Number of Submissions Received X 100): 30.66%**
- **Average number of reviews per paper: Two**
- **Total number of reviewers involved:51**
- **Any additional info on review process: NO**
- **Contact person for queries:** Dr. Girraj Sharm: girrajsharma.ece@jecrc.ac.in

  Dr. Ashish Kumar: ashishkumar.ece@jecrc.ac.in

X-Ray Determination of Debye Temperature and Microhardness of Some HCP Elements Re, Os and Tl

Purushotham Endla[1]

[1]Department of Physics, School of Sciences, SR University, Ananthasagar, Warangal Urban - 506371, Telangana, India.
E-mail: [1]psm45456@gmail.com

Abstract. The studies on the microhardness have been calculated the hexagonal close packed metals rhenium (Re), osmium (Os) and thallium (Tl). In this paper, the Vickers's pyramidal indenter and the true hardness values has been evaluated. Debye temperatures of hexagonal close packed (HCP) of these three metals have been calculated by using X- ray diffraction intensities pattern with Philips 3020 powder diffractometer fixed with a proportional counter using filtered CuKα radiation; all the values have been evaluated at room temperature In this present work the calculated Debye temperature values θ_H 338 K, 426 K and 74 for Re, Os and Tl respectively. The calculated Debye temperatures of these hcp metals have been estimated from the hardness H values 135, 400, 27 kg/mm^2 and are compared with those obtained from Debye temperature of specific heats (θ_D), Debye temperature of elastic constants (θ_E) and Debye temperature of X-ray intensity measurements (θ_M).

Keywords: X-**ray diffraction, hardness, Debye temperatures, hcp element**

1. Introduction

The hardness of the metal is an significant crystal property. The resistance obtainable by the lattice and its motion of dislocations. The crystal chemical armed forces in a crystal resist motion of dislocations as in it involves the dislocation of atoms/elements. This resistance is known as the intrinsic hardness of a crystal/solid. This crystal microhardness can be correlated by means the strength of interatomic-binding in the crystals.

The previous researchers Abrahams and Hsu [1], Purushotham and Gopikrishna [2, 3] and Denge [4] reported the thermal properties values of some metals and alloys. Mechanical Properties of pure metals and some alloys [5, 6] developed as potential engineering production [7] and thermal, mechanical properties and their applications [8].

2. Experimental

The samples Re, Os and Tl has been obtained from the commercial basis and available in the form of tiny ingots. A fine filing has been obtained in the form of the ingots through facilitate of a jeweller's file. The filing has been carried out slowly. The filings have passed through a 325-mesh screen. The X-ray Diffraction dimensions have been made with a diffractometer built-in with proportional counter. The X-ray tube was operated at 25 mA and 40 kV and. The XRD patterns of samples have been given in Figure1. The procedure for sample groundwork, have been given in earlier papers [9, 11].

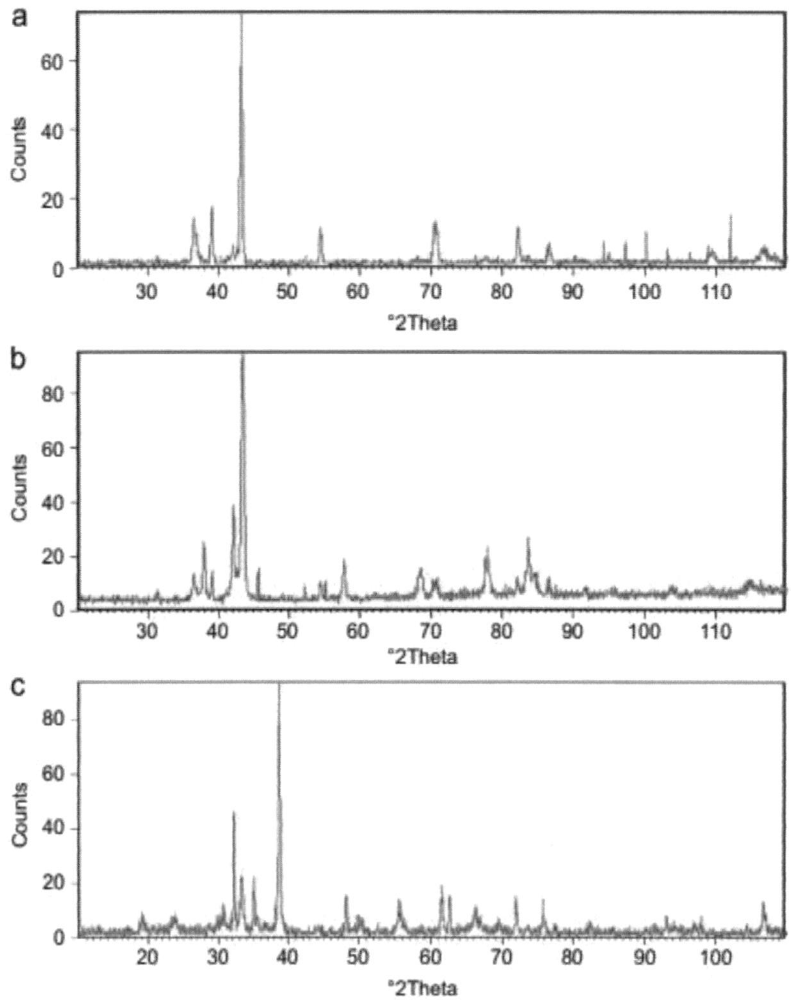

Fig.1. The X-ray patterns of a) Rhenium, b) Osmium and c) Thallium respectively.

3. Characterization Techniques

3.1 Debye model

The Debye model has been applied to examine the temperature variation of several physical properties of solids. One of the earliest applications of the Debye model was in the study of the effect of temperature on X-ray intensities. The treatment of this problem at the hands of Debye himself, led to the formulation of the Debye-Waller factor. The Debye-Waller factor accounts for the reduction in the intensities of the diffracted X-rays with temperature. Its primary use was in correcting the intensities of X-rays diffracted from crystals. However, very soon the relation between the Debye Waller-factor and other crystal properties was identified. Thus, the Debye-Waller factor can be used to calculate the amplitudes of vibration and the Debye temperatures. The Debye-Waller factor can also be theoretically evaluated from the lattice vibration spectrum. A comparison of the theoretically calculated value with the experimentally determined value can be used as a check on the model assumed in the calculation. Similarly, the Debye temperature can be correlated with the bond strength, the melting point and diffusion parameter. Thus, the value of Debye-temperature is no longer parameter useful only for the correction of intensities or for explaining the specific heats; they have emerged as independent solid state parameter.

3.2 Methods of determining Debye temperatures

The various experimental methods from which Debye temperatures can be determined. These are,

1. θ_E from elastic constants
2. θ_R from infrared frequencies
3. θ_ψ from compressibility
4. θ_{Tm} from melting points
5. θ_{ER} from electrical resistivity
6. θ_M from X-ray diffraction

The present thesis mainly deals with the experimental determination of mean amplitudes of vibration, Debye Waller factors and also Debye temperature values of some hcp metals, binary alloys with rhombohedral structure, ternary alloys with hcp structure and nano semi-conductors from the measurement of X-ray intensities. In the following section the Debye model, the concept of the Debye temperature and Debye-Waller theory are discussed. This is followed by a review of earlier work on the Debye-Waller factors and also Debye temperature values of metals, alloys and nanoparticles. The chapter concludes with a statement of the scope of the present investigation.

4. Results and discussion

Micro-hardness values have been carried/approved out by a micro-hardness tester fixed/settled with the Vickers's pyramidal indenter. The micro-hardness shows load dependence on enormously low load values and becomes self-governing of load. The load independence starts around 50gms. In order to find/discover load self-governing standards dimensions on all the three samples have been approved out at different loads up to 120 grams. The micro-hardness (H) is calculated using the equation (1)

$$H = 1.854\ P/d^2 \tag{1}$$

Where, P be the load (in grams) and d be the diagonal length (in microns). The micro-hardness falls as load increases and beyond 50gms the load dependence is negligible. In all the three cases the micro-hardness value at 120gms was taken as the accurate microhardness.

The Debye temperature of a solid can be calculated from the equation (2)

$$\theta_\psi = CV^{1/6}M^{-1/2}\psi^{-1/2} \tag{2}$$

Plendl [12] have been calculated micro-hardness (H) of a crystal, now substituting for ψ in equation 2, the following relation is calculated for Debye temperature (θ_H) in terms of hardness by using equation (3)

$$\theta_H = C'V^{1/6}\ M^{-1/2}\ H^{1/2} \tag{3}$$

Where, C' is a new proportionality constant. Seigal [13], Gopi Krishna and Kishan Rao [14] and Shankarnarayana [15] established the above equation for several hcp metals.

The above equation (3) represents the between Debye temperature (θ) and micro-hardness (H) has been calculated the values on the Re, Os and Tl by resources of the obtainable micro-hardness data in (Hand book of the Physicochemical properties of the elements, 1968) and the obtainable data on thermal properties like Debye temperatures have been calculated in the investigation by the X-ray method and also compared with the specific heat and elastic constant values, these values have been agreed well with each other. Used for this reason the constant C′ has been chosen to be 5×10^{-7}. The value of the constant is used for measuring the Debye temperature of all the present work Re, Os and Tl metals from the microhardness information. All the values have been summarized in Table 1. Gschneidner [16] have been considered Debye temperature from specific heats and elastic constant, these values also included in Table1.

Table 1. Values of Debye temperatures calculated from Eq.(3), specific heats, elastic constants and X-ray diffraction intensities for Re, Os and Tl.

Metal	H (kg/mm^2)	θ_H (K) Eq. (3)	θ_D (K)	θ_E (K)	θ_M (K) (Present work)	E_f (eV)
Re	135	388	429	421	413	5.00
Os	400	426	467	431	443	5.59
Tl	27	74	…	…	88	0.39

θ_H - From hardness values (calculated)
θ_D - From specific heats (Gschneidner 1964)
θ_E - From elastic constants (Gschneidner 1964)
θ_M - From X-ray intensities (Present work)

Experimental values of formation energies of metals may be obtained from quenching studies. A measurement of the quenched-in electrical resistivity as a function of quench temperature yields a value for the formation energy as residual electrical resistivity is proportional to the total number of defects frozen in by the quench. Experimental values of the formation energy are available for several metals (Peterson, 1968). The values of formation energies from experiment are not known for many hcp metals and it is desirable to obtain the formation energy from any other known physical quantity. An attempt has been made here to obtain the information from temperatures.

Mukherjee [17] empirically obtained the relation

$$\theta = C\left(\frac{E_f}{MV^{\frac{2}{3}}}\right)^{\frac{1}{2}} \tag{4}$$

Where θ is the Debye temperature, E_f the vacancy formation energy, M and V are the mass and volume per atom of the crystal and C is a constant which is the same for all metals. From his empirical calculation Mukherjee suggested a value of 34 for the constant C.
Glyde [18] relation is given by

$$E_f = A\left(\frac{k}{\hbar}\right)^2 M\,\theta^2\,a^2 \qquad (5)$$

Where a is the inter-planar spacing, $A = 1.17 \times 10^{-2}$, M be the molecular weight and h and k are the Planck and the Boltzmann constants respectively. Purushotham [19, 25] have been used the above equation (5) and calculated vacancy formation energy for several metals and alloys. The calculated values of vacancy formation energies for Re, Os and Tl are also included in Table 1.

Conclusions

In the resent study, the micro-hardness has been calculated the three hexagonal close packed metals Using Vickers's pyramidal indenter and the true hardness has been estimated. In this present work the calculated Debye temperature values 338 K for Re, 426 K for Os and 74 K for Tl metals. Debye temperatures values of these three hcp metals were measured from the hardness values 135, 400, 27 kg/mm^2 and are compared with those obtained from Debye temperature of specific heats (θ_D), Debye temperature of elastic constants (θ_E) and Debye temperature of X-ray intensity measurements (θ_M). The microhardness of three hcp metals been calculated. The thermal properties of these hcp metals have been measured using the present micro-hardness values and also compared with Debye temperature values calculated from specific heats, elastic constants and X-ray diffraction studies.

References

1. Abrahams, S.C. and Hsu, F.S.L.: J. Chem. Phys. 63, 1163 (1975).
2. Purushotham, E., Gopi Krishna, N.: Physica B, Volume 405, Issue 16, Pages, 3308-3311 (2010)
3. Purushotham, E., Gopi Krishna, N.: Indian Journal of Physics, Volume 84, Issue 7, Pages, 887-893, (2010)
4. Deng, X.H., Lu, W., Hu, Y.M. and Gu, H.S.: Physica B 404, 1218-1221 (2009).
5. Ashok Kumar, V.,Sammaiah, P.: Science Direct Materials Today:Proceedings1-7, (2017).
6. SrikanthRao, P., Manoj Kumar, D., Gopikrishna, J.: Science Direct Materials Today, Vol.5, issue.1 (1) 1264-1270 (2018)
7. Shiva Krishna, Ch., ManojKumar, J.: International Journal of Materials Science,Vol.12, issue.4, 627-633 (2017).
8. SatishKumar, P., Sastry, CH.S.R., Devaraju, Ch.: Materials Today Proceedings Elsevier,Vol.4, issue.2, 330-335 (2017
9. Purushotham, E., Gopi Krishna, N.: Bulletin of Materials Science (springer) Vol 36,Issue 6, Pages 973-976 (2013)
10. Purushotham, E,: Journal of engineering science and technology review, Volume 7, Issue 1, Pages 41- 44 (2014)
11. Purushotham, E.: Rasayan Journal of Chemistry, Volume 12, Issue 4, Pages 1676-1680 (2019).
12. Plendl, J.N., Gielisse, P.J.: Journal of Applied Optics, 4, 853 (1965).
13. Seigal, E.: Flammarian Sciences, Paris (1978).
14. Gopikrishna, N., Kishan Rao, K.: Bulletin of Pure and Applied Science, 23, 97-101 (2004).
15. Shankar Narayana, M.: Doctor of philosophy thesis,) (2007).
16. Gschneidner, K.A. Jr.: Solid State Phys. 16, 276-426 (1964).
17. Mukherjee, K.: Phil. Mag. 12, 915 (1965).
18. Glyde, H.R.: Journal of Physics and Chemistry Solids (GB), 28, 2061 (1967).
19. Purushotham, E,: Journal of engineering science and technology review, Volume 7, Issue 1, Pages 41- 44(2014).
20. Purushotham, E.: Rasayan Journal of Chemistry, Volume 12, Issue 4, Pages 1676-1680 (2019).
21. Purushotham E and Gopi Krishna N.: Bulletin of Material Science, 37 773(2014)
22. Purushotham E and Gopi Krishna N.: Physica B, 405 3308 (2010).
23. Purushotham E and Gopi Krishna N.: Indian Journal of Physics, 84 887(2010)
24. Purushotham E and Gopi Krishna N.: Bulletin of Material Science, 36 973 (2013).
25. Purushotham E.: Chemistry and Material Reviews, 7, 1(2015).

IOP Conf. Series: Materials Science and Engineering 1119 (2021) 012002 doi:10.1088/1757-899X/1119/1/012002

Optimized Scale Invariant HOG Descriptors for Object and Human Detection

Akila K[1], Pavithra P[2]

[1,2]Assistant Professor, R.M.K College of Engineering &Technology, Chennai, India
[2]pavithracse@rmkcet.ac.in

Abstract

This article presents a scale variation approach to identify objects and humans in video sequences using histogram of gradient descriptor. A significant restriction in HOG descriptors is its variations with scale changes and illumination changes, as is frequently the considered case. We recommend unique SIO-HOG descriptors that are figured to be invariant to scale changes. The system associates the benefits of adoptive bin selections and sample resizing in the object recognition process. We analyze the effect of PCA transform based feature selection process on object detection performance, ultimately the finite scale range, adoptive orientation binning in non-overlapping descriptors is all main thing for nominal detection rate. HOG feature vector over complete search window is computationally more exclusive, to acquire more precise set of features with finite Euclidean distance to classify them using KNN classifier. This new approach provides near-perfect ways of separating humans from other objects. The whole object detection system was assessed on a few test samples from real-world data sets and compared beside a publicly available pedestrian detection data base.

Keywords: scale-invariance, HOG descriptor, object detection, PCA transform, spatial features, classification

1. Introduction

Detecting objects in images or vision input is computationally exhaustive tasks since massive quantity of pixels needs to be processed and complex arithmetic operations are considered. Even then, the outcome is still more amount of features that prerequisite to be calculated due to its variable appearance, scale changes and the pose variations. To incorporate robust feature sets like edge orientation histograms [1], SIFT descriptors [2] are computed on a consistently spaced cells to overcome all these trials. We briefly discuss previous work on object detection in section 2, give a detailed overview of our HOG method section 3, describe the detailed description and experimental evaluation of object and human detection process in section 5 and concluded in section 6.

2. Previous Work

The improvements in classifier routines and the increasing accessibility of large range of feature sets to encode local information of objects offers quite a few ways to expand the detection precision. In [3] Haar-like wavelet feature set and space-time differences structures were used to figure an effective region-based detection for moving person detector. In [4] use the mixtures of orientation histograms with threshold gradient to compute feature vector separately for each parts of the object. HOG features have established its discernment for the many object description and detection. HOG feature sets are not only used to describe the appearance of the object and it is also used to reveal the object shape. This creates them elect and actual technique for object detection. However, it has been used in a few object detection mechanisms it has the incapability to detect objects in various poses. This problem is resolved using discriminatively trained models with the assistance of multi-class SVMs [5]. Survey on Illumination Condition of Video / Image under Heterogeneous Environments for Enhancement. [6] To preserve the uniqueness, sustaining the original scale invariant property mirror invariant descriptors are framed according to the magnitude beneath flip and rotation. [7] Moreover scale invariant based on shape context, other local feature sets such as Continuous wavelets [8] and SIFT [9] have been used as key point-based detection methodologies.

3. HOG descriptor

Histograms of Oriented Gradients are high-level features computed based on image gradient of decomposed patches. The HOG feature vector constructed from each object are used to assess the detection process with limited discrimination. Feature vectors are consequently train the classifier for automated object detection. It has been used in utmost object detection by evaluating well-normalized image gradient orientations in the past decade [10]. Basically, HOG descriptor is estimated based on two dissimilar geometries, rectangular HOG with square spatial cells, and circular HOG partitioning cells in log-polar fashion.

3.1 R-HOG

R-HOG blocks have many likeness measures with scale invariant descriptors but adaptive in nature to local image conditions since RHOG is computed at a single scale without dominant orientation alignment, whereas SIFT's computed at a scale-invariant key points, with their dominant orientations. [11] RHOG descriptors typically consist of multiple block types with different cell and block sizes to advance the performance.

3.2. C-HOG

Circular block (C-HOG) descriptors are frequently look like shape frameworks but feature vector comprises gradient-weighted orientation cells instead of a single edge orientation cell. CHOG descriptors are having radial bins to advance the performance in the form of centre-surround coding.

3.3. Spatial / Orientation Binning

The usage of orientation histograms for object detection has many pioneers [12], but it spreads constancy only when it is combined with local spatial information and normalization. Independently HOG feature vector calculates edge orientation histogram and are assembled into orientation bins over object regions i.e spatial cells. Cells are rectangular blocks, and the orientation bins are equivalently spread out over $0^\circ - 180^\circ$ as an unsigned gradient with step size of 20° to 40°. To reduce spatial redundancy, image search window is divided into non overlapping blocks since orientation bins are interpolated.

4. Object Recognition

The over-all architecture of the object detection system is shown in Figure 1. Subsequent the macro block conversion paradigm, [13] the HOG extraction process has been characterised into two phases: bin selection based on intensity variation and re sampling based on appearance changes. The workflow is as follows: The primary stage, the precomputation stage, generates bins and sample size, i.e., rectangular regions and angular bins which generate HOG feature vectors. These extracted HOG get into the next level, the feature selection stage, which analyzes feature sets and selects feature vectors with great variance in a more systematic way. [14] It supports the existence of a most significant feature

IOP Conf. Series: Materials Science and Engineering 1119 (2021) 012002 doi:10.1088/1757-899X/1119/1/012002

and rejects the redundant sets. To end with classification system output relates bounding boxes on object detected windows with described size.

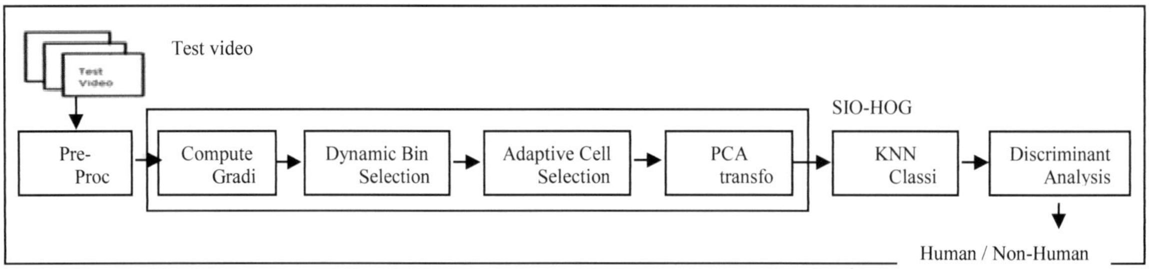

Figure 1. General architecture of the object detection system

4.1. Precomputation stage

In the HOG generation stage, we use an adoptive bin selector and sampling size for robust object detection over poor illumination and fast scale changes due to the two subsequent facts (for details on feature extraction [E]):

- **Simple feature set**: HOG features provide a very basic set of features that is computed efficiently from image patches.
- **Bin size:** HOG feature vector magnitudes are directly depending on input image pixel values. Hence changes in the intensity level will change the HOG description. Feature vector is more sensitive when the number of bins is high.

4.2 Feature selection stage

To advance a hypothesis HOG feature set that detects as many objects as possible; the detection rate should be kept high, even if the HOG vectors lied in false positives. For that reason, we carried out principal component analyzes (PCA) to select dominant features, using hierarchical correlation matrix followed by variance computation. This also helps to detect the smallest objects even with the wide range of different poses and appearances.

Pseudo code:

```
Step 1:  Frame conversion
Step 2:  For i= 1 to Total number of frames
              for each frame I
Step 3: select bin & cell size.
                    Compute HOG gradient model

Step 4: PCA transform analyzes.
                    for m=1: no. of HOG features
                       find variance & correlation matrix.
                         If correlation level > integer bound
                              Include parameter SET!
                         end
Step 5: carry out discriminative analyzes.
                         set orientation level.
                         check symmetric level.
Step 6: If detected HOG model = = Human HOG prescription
                                   Matching count++.
                            end
                            if Matching count > parameter SET
Assert HOG optimization and input parameter set evaluations.
Step 7: End
```

In the last stage, the hypotheses are confirmed with computationally efficient classification mechanism. It uses a different set of HOG features extracted from previous stage. Whereas the size of feature vector is slower, the classification performance of KNN classifier is more accurate, since the number of false positives is low.

The HOG feature vectors are extracted over a search window with the size of 64 × 128 pixels and 8x8 cell size; this allows capturing detailed local information. 10 positive samples were taken from the PASCAL VOC 2007 dataset and trade-off between window sizes over detection rate is analysed. The object detector analyses over various resolution images, Scales range from 68×56 to 128×64 pixels, possible cell sizes. As shown in Figure 2 16x16 cell size gives poor detection rate even with larger window size. Since the object samples are quite small, the HOG descriptor is limited to minor details. Thus, we found that low resolution object samples have little effect on the detection performance when the cell size is small.

4.4 Human detection

An image, I, is first partitioned into multiple non overlapping 8 × 8 patches Pi, i = 1,…, N . For each patch, a 9-bin HOG descriptor with the quantization angles j × 20°, j = 0...8 is used to signify its local spatial information's. Though, HOG descriptor deficiencies the symmetric properties as shown in Figure 3, and consequently we distinct the symmetric relationships of HOG descriptors. Figure 4 illustrations that large cell size of an object holds a peak bin angles and its corresponding HOG approximates the gradient direction.

As cited above, image patches be appropriate to the human spatial regions form a symmetric HOG vectors. Thus, the symmetric HOG feature vector can be defined as:

$$M= \{(X_i,X_j) \mid X_i \in T1 \; X_j \in T2$$

where T1- Train sample, T2- Test sample

Note that the HOG feature vectors need to be large enough to preserve most of the potential symmetric features. Thus, the computational complexity to execute the human detection is not high. The local similarity measurement for $\{X_i,X_j\}$ on the HOG (r, θ) space P: $P(r,\theta)=P(r,\theta)+Sim(X_i,X_j)$

4.5. Feature extraction

The extracted features are based on histograms of local gradients (HOG) that are computed from the Non- overlapping image patches in the search window. For each cell size, its histogram values describe the gradient magnitude in particular directions. As stated, former histograms with 9 bins (ranging from 0 ◦ to 180◦) seem to be select for object detection process. Since the image patches are rectangular and the histogram bins are efficiently computed to build HOG feature vector: To work out integrated samples gradients are extracted from non-overlapping blocks to allows saving time for object detection process.

5. Experimental results

Here KNN classifier is used to assess the performance of object detection by means of gradient-based HOG descriptor with adoptive bin selection scheme. Meanwhile prior information's are not existing about objects for choosing the class with the major probability. Generally, there are some notable outcomes in object detection rate. The fact that the scale invariant HOG greatly out-performs existing HOG model and that any important modifications in the intensity level before compute HOG gradients harms the true positive samples highlights that ample of the available spatial orientation information is from the fine scales, and the confidence of dropping the sensitivity to illumination and scale changes is increased. Here HOG gradients are calculated at the finest image scale, rectified for adoptive orientation binning, and invariant to the blurred spatial object region. On the other hand, at least for human detection, it pays more consideration to asymmetric nature of orientation rather fine symmetric vectors.

Table 1: Recognition Rates of Dense and Custom Descriptor

DATA set-PASCAL VOC 2007 data base	Block decomposition	Eigenvectors selected.
EHOG model-PCACarl Vondricket a.2013)	Overlapping model	28
SIO-HOG model -PCA	Non overlapping model	22

Table 2: Average detection rate performance of proposed HOG class in the PASCAL VOC 2007 dataset.

DATA set-PASCAL VOC 2007 data base	False detection rate	Average detection rate performance
EHOG model (Carl Vondricket a.2013)	13%	16 classes
SIO-HOG model	10%	19 classes

Table 3: Human detection rate analyzes in the INRIA pedestrian dataset.

DATA set-INRIA pedestrian data base	False detection rate	positive detection rate
HOG model (M. Kachouane et a.2012)	17%	86%
SIO-HOG model	12%	92%

IOP Conf. Series: Materials Science and Engineering 1119 (2021) 012002 doi:10.1088/1757-899X/1119/1/012002

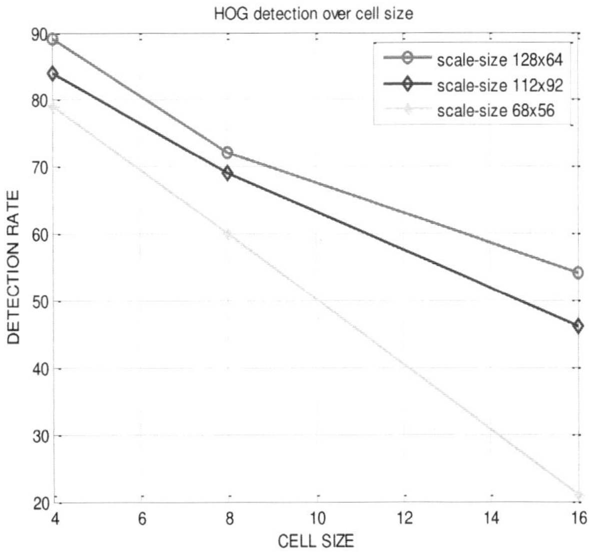

Figure 2. Detection rate analyzes of HOG over Scales range from 68×56 to 128×64 pixels.

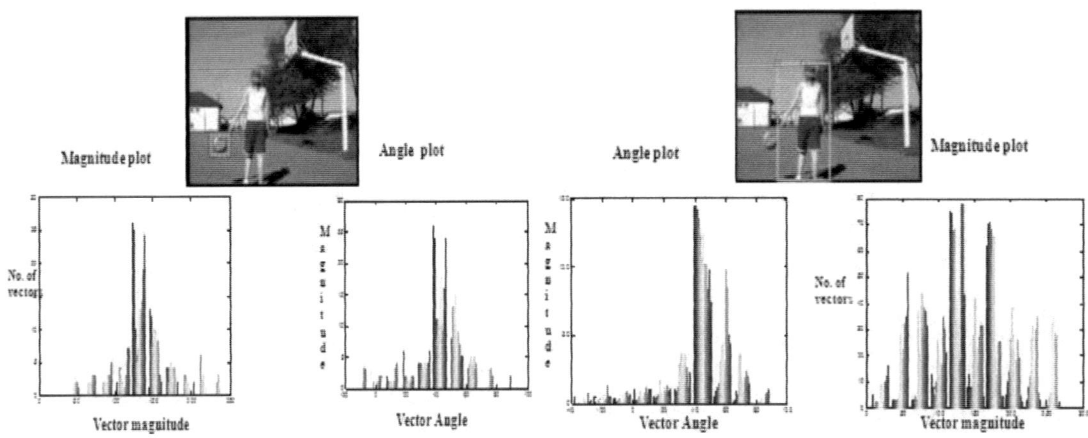

Figure 3. Symmetric properties of HOG model for human and object discrimination

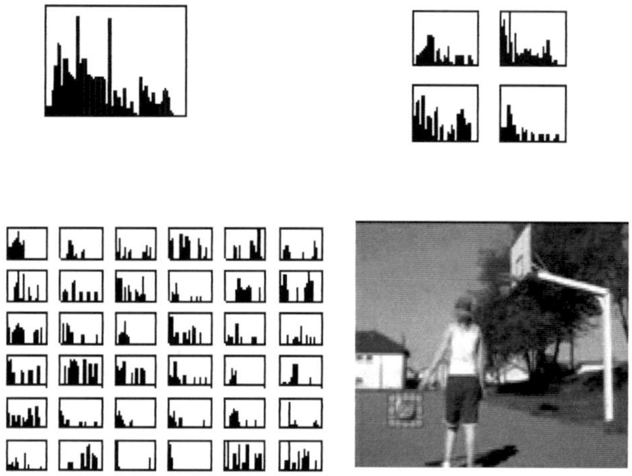

Figure 4 . Histogram bins of variable cell size of an object

6. Conclusion

We proposed novel SIO-HOG descriptors, which is mined by transforming HOG descriptors to be invariant to scale variation and illumination changes. Different appearance variation and illumination changes have been estimated to demonstrate the robustness of the HOG descriptor configuration. We analyze the performance metrics of our HOG model in both object and human detection and proved it is superior to the existing HOG descriptors through experiment with the PASCAL VOC 2007 dataset and INRIA pedestrian data sets. Furthermore, we exhibited the efficiency of effective feature reduction method. Meanwhile we analyzed the object detection and recognition rates with real time world data we substantiated that the presented object detection system marks an enhancement compared to systems solely based on HOG orientation with constant bin selection. Additional researches also will emphasis on the implementation of human action recognition system to show robustness of proposed HOG model over human detection rate per frame.

References

[1] Freeman W T and. Roth M 1995 Orientation histograms for hand gesture recognition. Intl. Workshop on Automatic Faceand Gesture-Recognition, IEEE Computer Society.

[2] Lowe D G 2004 Distinctive image features from scale-invariant key points. IJCV.

[3] Viola, M. Jones J. and Snow D 2003 Detecting pedestrians using patterns of motion and appearance. pp.734–741.

[4] Mikolajczyk K Schmid C Zisserman A 2004 Human detection based on a probabilisticassembly of robust part Detectors. 8th ECCV, pp. 69– 81.

[5] K.Akila, S.Chitrakala, S.Vaishnavi 2016 Survey on Illumination Condition of VideoImage under Heterogeneous Environments for Enhancement ICACCS 2016 IEEE

[6] Felzenszwalb P F Girshick R B McAllester D 2010 Object detection with discriminativelytrained part based models IEEE on Pattern Analysis and Machine Intelligence.

[7] Xiaojie Guo and Xiaochun Cao 2012 MIFT: A framework for feature descriptors to be mirror reflection invariant Image and Vision Computing. Pp. 546– 556.

[8] Reddy Vijaya Kumar Kiran Kumar,Object Detection by 2-D Continuous Wavelet Transform ComputationalScience and Computational Intelligence IEEE.

[9] Zhao Wan-Lei Chong-Wah Ngo 2013 Flip-invariant SIFT for copy and object detection IEEE Transactions on Image Processing pp. 980-991.

[10] Mikolajczyk K Schmid. C 2004 Scale and affine invariant interest point detectors IJCV pp.63 86.
[11] Kanezaki Asako Yusuke Mukuta, and Tatsuya Harada 2014 Mirror reflection invariant HOG descriptors for object detection Image Processing ICIP IEEE Conference.
[12] Pedersoli Marco TinneTuytelaars 2014 A scalable 3d hog model for fast object detection and viewpoint estimation 3D Vision 2nd International Conference -IEEE.
[13] Chayeb A 2014 HOG based multi-object detection for urban navigation IntelligentTransportation Systems
[14] Chen Lu Panfeng Huang, and Jia Cai 2015 Object detection for non-cooperative targets using HOG-based proposals Robotics and Biomimetics.
[15] Akila K Chitrakala S 2014 Discriminative Human Action Recognition using HOI Descriptor and Key Poses ICSEMR
[16] Singh Shekhar Gupta S C 2016 Human object detection by HoG features PDGC IEEE conference
[17] Akila K Chitrakala S 2019 Highly refined human action recognition model to handleintraclass variability & interclass similarity International Journal of Multimedia

Design and Implementation of Security Device for Patient Health Monitoring Systems

A.K. Sahu[1], **N. K. Misra**[2*], **D. Kumar**[3]
[1,2,3]Department of Electronics and Communication Engineering, Bharat Institute of Engineering and Technology, Hyderabad, Telangana, India
Email Id: [2*]neeraj.mishra3@gmail.com

Abstract. In continuous upgrading world, humans believe in their self-worth. They have participation in every sector of life, but lives have become so vulnerable these days that the safety and security of their lives are one of the burning questions about this pandemic corona virus disease. Considering all incidents and violation of rules do not spread against humanity this idea of a smart wristband safety device aligned with GPS and GSM modules, with temperature and pulse sensors came into consideration. During dangerous situations user just needs to press the SOS button fitted on the wristband, or the sensors will sense an increase in temperature or pulse rate and then automatically the message of user location tracking via GPS will be sent to the registered numbers through GSM. The main objective is for the device to be light weight and place the SOS switch in an easy accessible region, with additional sensors leaving no worse case possible.

Keywords: Embedded System, health information processing, Security, Device and circuits, sensors.

1. Introduction

Health has become a major issue in the world now in COVID-19 pandemic. The spreading rate against humanity in the country has only risen to a great extent. India ranks 3th in the COVID-19pandemic [1]. There are many scenarios where COVID-19 case is not being reported by the human, which would add up to a much higher rate. Human kind, of every perspective needs to stand common on the grounds of individual safety by not causing harm and panic to other citizens. There are fewer laws on safety and security of human in the health care section [2]. Likewise, there is no much to be done after the COVID-19 has been committed. Every human contribute to the growth and development of the country, but live in fear. At times there are situations where there will be no one for the patient to help.This work designed a security device that has GSM modules included in it. Temperature, pulse rate sensors and SOS switch are the key subjects depending on which an alert message will be sent to registers mobile number.

2. Motivation and background of work

The challenging situations facing by everyone now-a-days gave motivation to come up with a security device to help everyone to do the work they liked to do. Devices existing in this context have been a source of motivation of our work [3]. Safety being provided to healthy people, before virus infection taking place is a better start to protect such kind. With rapid increase technicality wise, our country stands in a better position in inventions. By this an embedded system resolving this can be made. Having the concern, many developers have come up with products including applications of COVID-19. The issue with using an app is the Wi-Fi that it requires which is not always available. Using only an SOS button, will also have few barriers as the person might not always be in a condition to press the button manually, and using only this feature is not quite helpful to the modern problem [4].This productdoes not exist in the market similar one.In similar one has sensors and also SOS button, but the main issue arising is only one mobile number can be registered and it is not always practical having a single source of help. Also the price is high and not reasonable [5]. Figure 1 shows Covid replication in which electron tomography to show that after disease of cells, MERS-CoV incites different interconnected to change film structures, including tangled layers and DMVs, which are signs of beta-CoV contaminations, just as DMSs, which have not recently been accounted for beta-CoVs. Besides, the creators found that another beta-CoV (murine hepatitis infection), just as a gamma-CoV (irresistible bronchitis infection (IBV)) and an individual from the variety alpha-CoV (human Covid 229E) prompted a similar layer structures in various cell lines [7] [8]. Figure 2 shows the temperature Pal brilliant thermometer [6] in clinic frameworks. Moura. (2020) et al. Discussed state of the craftsmanship, research issues and future patterns, with an attention on resilience [9]. Abdelnasser et al.(2020) talked about COVID-19. An update for muscular specialist [10]. Chaudhari et al. (2020) discussed the part of the web of things (IoT) in pandemic Coronavirus condition [11].

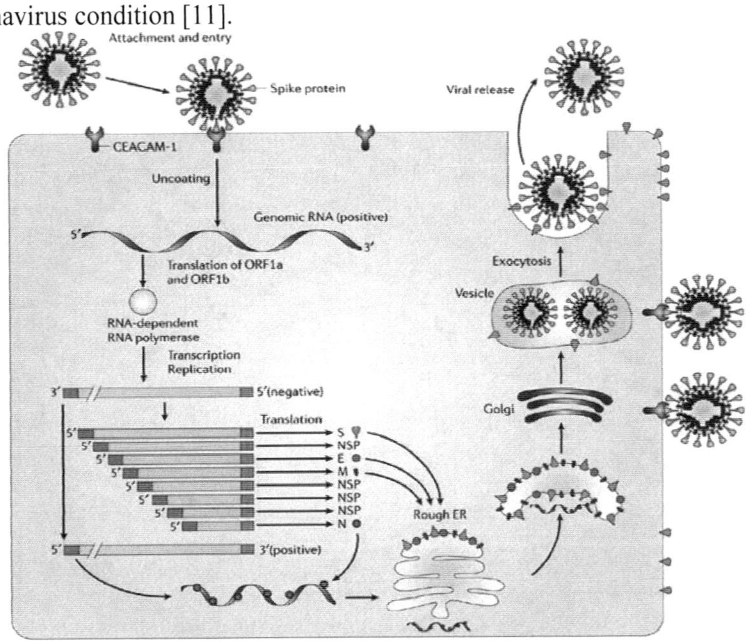

Figure 1. Coronavirus Replication [7]

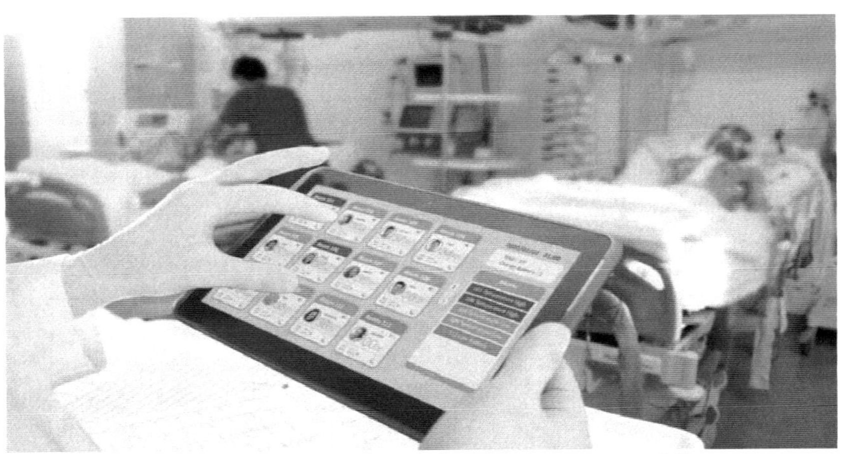

Figure 2. Temperature Pal smart thermometer [6]

3. Hardware Requirement for Implementation

This framework capacities with 4 distinct modules, for example, SIM800L MODULE (Location Tracking Module), temperature sensor, heart beat sensor (Sending module), SOS Button. The determination of parts should be with the end goal that the size of the gadget should be conservative and the reaction time taken by every single module should be less for fast yield. This significant arrangement work quality is estimated as far as thought offactors, including, morals, cost, type and guidelines. In second module as Node MCU utilization of Microcontroller: ATMEGA 328P RISC: 32 BIT. In the third module as SIM 800L, in this GSM and GPS module are independently accessible on the lookout. As space imperative is a significant issue in planning a minimal gadget, SIM 808L is the joined of GSM and GPS module. In temperature sensor modules we are utilizing two sensor one is LM – 35 (temperature sensor) and another is an infrared finger cut sensor. For controlling we are utilizing SOS key will permit you to send a predefined custom SMS message, including your present area, to up to 3 predefined contacts through GPRS associated with the gadget. The gadget is planned like a convenient cudgel. The Node MCU requires a flexibly voltage of 3.3-5V. The SIM 808, SOS crisis button, LM-35 (temperature sensor) and infrared finger cut sensor (heartbeat sensor) interfaced to the Node MCU. Circuits and their operations in equal information moves and sequential correspondence move utilizes for 1 correspondence which makes the information move less expensive, yet in addition makes it workable for two PCs situated in two distinct urban areas to impart over phone. Sequential information correspondence is as non-concurrent and coordinated. The synchronization strategy moves information at a time while the offbeat exchanges a solitary byte at a time. These chips are generally alluded to as UART and USART. Figure 3 shows the circuit diagram of Coronavirus gadget for quiet well being checking framework.

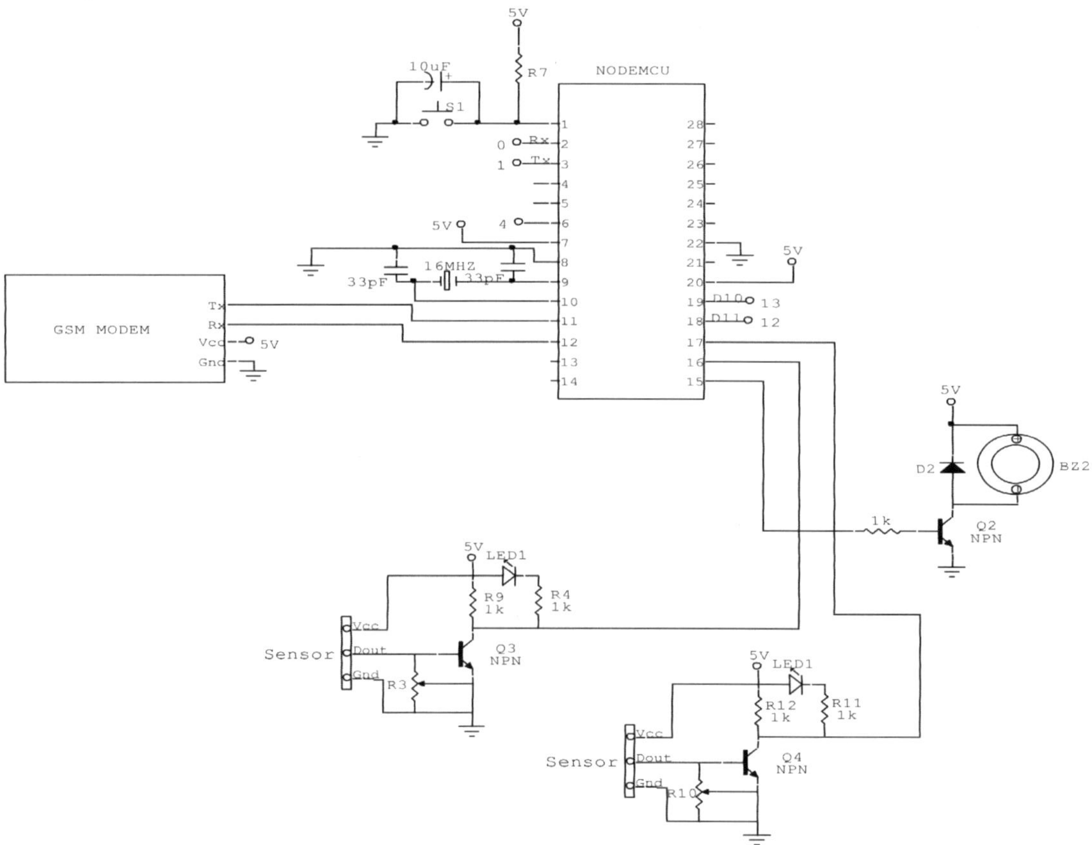

Figure 3. Circuit design of COVID-19 Device for Patient Health Monitoring System

4. Circuits and their operations

This security device which is a wrist band has an SOS emergency button, a temperature sensor to detect the body temperature using a sensor called LM-35, a heartbeat sensor to detect the pulse of the person wearing this wristband which is an infrared finger clip sensor and a module named SIM 800L which is the combination of GPS (Global positioning system) and GSM (Global system for mobile) all connected to a Microcontroller used NODE MCU. When the healthy person is suspected to be in danger, the SOS emergency button is pressed once, the microcontroller initializes and using GPS the location of the healthy person is tracked (values of longitudes, latitudes) and then the location is sent to the registered mobile numbers saved using GSM When the healthy personas in suspect of danger and because of situations the healthy person wasn't able to press the SOS emergency button a set of sensors are connected to be helpful for detection. When the person is unconscious, the temperature sensor(LM-35) and the heartbeat sensor(infrared finger clip) detects the changes in the body temperature and changes in the pulse of the victim, this device still sends the location of the healthy person using GPS(values of longitudes and latitudes) to the registered emergency mobile numbers using GSM. The registered mobile numbers will receive the values of Longitudes and latitudes of the location of the healthy person and can rescue them before the incident takes place. Figure 4 showsa flow diagram of covid-19 device for patient health monitoring system in architecture. Figure 5 shows hardware onthe security device for patient health monitoring system. Figure 6 shows mobile output display of security device for patient health monitoring system.

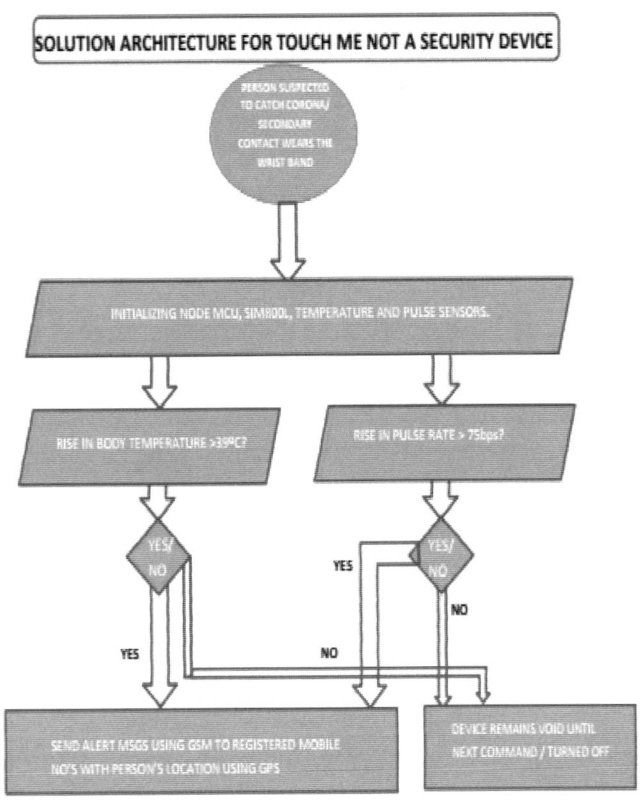

Figure 4. Flow diagram of COVID-19 Device for Patient Health Monitoring System

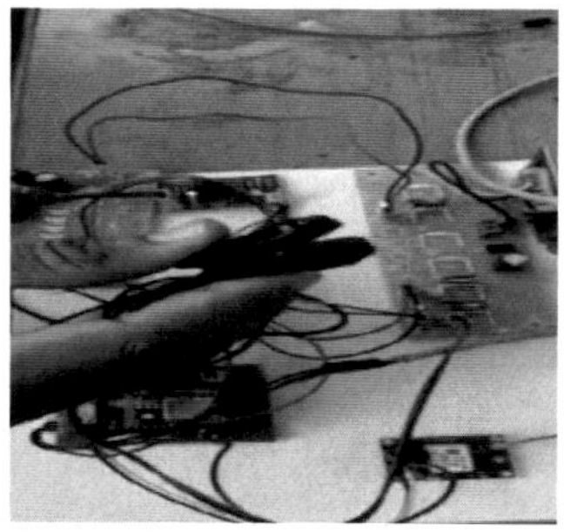

Figure 5. Hardware of Security device for Patient Health Monitoring System

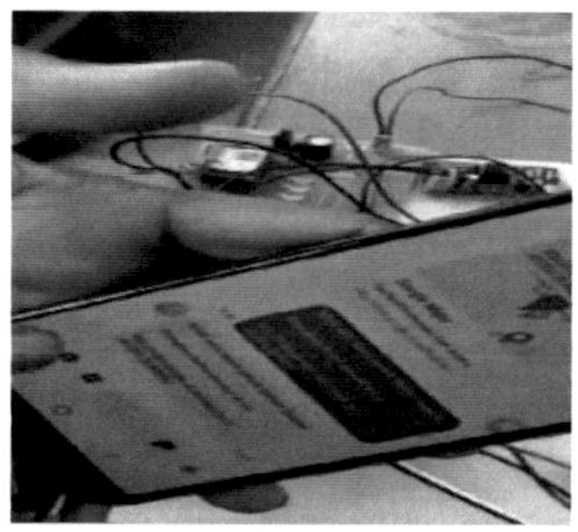

Figure 6. Mobile Output display of Security device For Patient Health Monitoring System

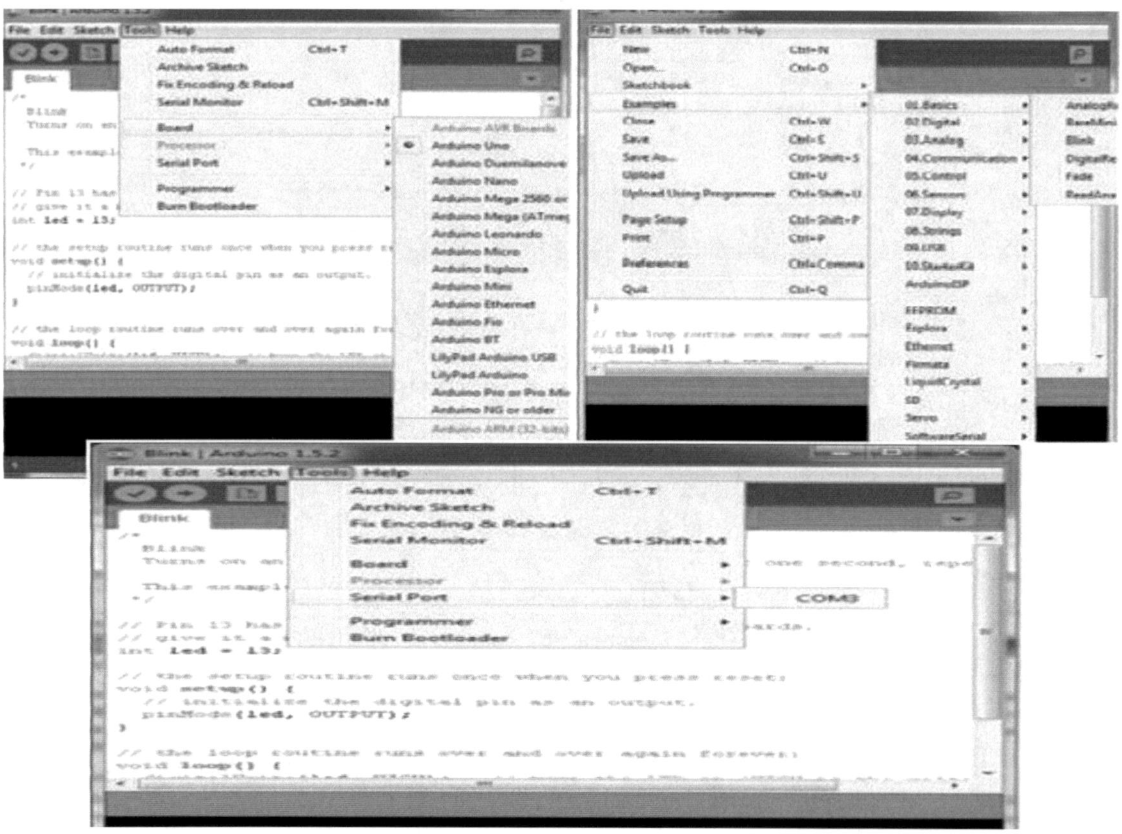

Figure 7. Software Development of Security device For Patient Health Monitoring System

After following the appropriate steps for your software install, we are now ready to test your first program with your Arduino board. Figure 7 shows the software development of security device for patient health monitoring system.
Comparative analysis of cost and performance: SOS button is used. It sends alert texts to receiver mobile numbers with the location. The devices send message, whereas here mobile app is being used. Sensors have not been used. This system is comparatively less cost compared to existing systems. The manual used to trigger the device is also possible in this system.

5. Conclusion

This framework is intended to help the solid individual in trouble circumstances to call for help just as ready the encompassing individuals. It has been created to make the security gadget more compact and agreeable. This framework was planned after reference to effectively actualized framework. Here measures are being made to conquer the blemishes in past plans and furthermore to decrease the force utilization. In future extension incorporates the ongoing execution of the proposed framework with the extra segments like a signal to caution the individuals close by, and a camera to catch pictures of the COVID-19 patient. Our future degree incorporates the continuous usage of the proposed framework with the extra segments like a ringer to alert the individuals close by, and a camera to catch pictures of the COVID 19 patient.

References

[1] Velayutham, R., Sabari, M., &Rajeswari, M. S. (2016, March). An innovative approach for women and children's security based location tracking system. IEEE International Conference on Circuit, Power and Computing Technologies (ICCPCT) (pp. 1-5).

[2] Harikiran, G. C., Menasinkai, K., &Shirol, S. (2016, March). Smart security solution for women based on Internet of Things (IOT). IEEE *International Conference on Electrical, Electronics, and Optimization Techniques (ICEEOT)* (pp. 3551-3554).

[3] Punjabi, S. K., Chaure, S., Ravale, U., & Reddy, D. (2018, November). Smart intelligent system for women and child security. *IEEE 9th Annual Information Technology, Electronics and Mobile Communication Conference (IEMCON)* (pp. 451-454).

[4] Bhilare, P., Mohite, A., Kamble, D., Makode, S., &Kahane, R. (2015). Women employee security system using GPS and GSM based vehicle tracking. *International journal for research in emerging science and technology*, *2*(1), 65-71.

[5] Vidyasagar, K., Balaji, G., & Reddy, K. N. (2015). RFID-GSM imparted School children Security System. In *Communications on Applied Electronics (CAE)* (Vol. 2, No. 2). Foundation of Computer Science FCS.

[6] Snijder, E. J., Limpens, R. W., De Wilde, A. H., De Jong, A. W., Zevenhoven-Dobbe, J. C., Maier, H. J., ... &Bárcena, M. (2020). A unifying structural and functional model of the coronavirus replication organelle: tracking down RNA synthesis. *Plos Biology*, *18*(6), e3000715.

[7] Moura, J., & Hutchison, D. (2020). Fog computing systems: State of the art, research issues and future trends, with a focus on resilience. *Journal of Network and Computer Applications*, 102784.

[8] Abdelnasser, M. K., Morsy, M., Osman, A. E., AbdelKawi, A. F., Ibrahim, M. F., Eisa, A., Abdelgawaad, A. S. (2020). COVID-19. An update for orthopedic surgeons. *SICOT-J*, *6*.

[9] Chaudhari, S. N., Mene, S. P., Bora, R. M., &Somavanshi, K. N. Role of Internet of Things (IOT) In Pandemic Covid-19 Condition.

Reversible Gate Mapping into QCA Explicit Cells Packed with Single Layer

N Pathak[1], **N K Misra**[2*], **B K Bhoi**[3], **S Kumar**[4]

[1,4]Department of Computer Science and Engineering, Maharishi University of Information Technology, Lucknow, Uttar Pradesh 226013, India
[2*]Department of Electronics and Communication Engineering, Bharat Institute of Engineering and Technology, Hyderabad, Telangana, India
[3]Department of Electronics and Telecommunication, Veer Surendra Sai University of Technology, Burla 768018, India

Email Id: [2]neeraj.mishra3@gmail.com

Abstract. MOS based circuits and computing devices are facing challenge related to energy dissipation, short channel effect and device density. The state of the art technology is driven towards Quantum cellular automata (QCA) computing with an emphasis on a high-speed and smaller area. QCA is a new computing technology that is made of quantum cell restraining two electrons and two dots. In this paper, a novel design of Fredkin, Toffoli, and Feynman gate layout is designed using the molecular QCA concept. Presented cell layout of newly circuit validity is verified by QCADesigner software. Further, a large number of articles are reviewed and comparison results are presented to show the optimal results such as cell count, majority gate count, bounded box area, and clock utilize. In this work, QCA and reversible logic combined together and the proposed designs have less bounded box area and delay, which is an essential part of the next-generation computing.

Keywords: Reversible Logic, QCA nanocomputing, Molecular QCA, Low power design

1. Introduction

According to Moore's law, the chip design industry has evolved spectacularly. Eventually, the chip size has been minimized to nanoscale and the processing speed has been increased immensely [1]. Now for the engineers, it is the trade-off between power dissipation and increased processing speed. To minimize the power dissipation different technologies are available like charge based recovery [2]. By going beyond conventional method energy dissipation can be minimized. For this reversible logic can be used as a solution. Rolf Landauer [3] manifested for a one-bit of digital logic signal during data erasing KTln2 joules of energy is dissipated. During computation in irreversible logic, some information are being erased so as to avoid its reversible logic circuits are used. Bennet demonstrated by using reversible logic circuit zero power dissipation can be possible [4]. A logic binary gates are known as reversible if its equal input and output in numbers an input to output mapping is Bijective [5]. Because of this property, no information loss occurs in the reversible circuit. The reversible circuit must have minimum garbage output [6]. In this article study and designs of high speed reversible gates

layout in QCA has been introduced. Results and discussion show a higher level of performance for the new layout in QCA are better than CMOS at high speed computing applications.
The major features of the proposed work are as follows:

- The reversible gate architecture is designed using Boolean expression and truth table which can be further used for designing QCA layouts.
- The parameter analysis of reversible gate is done by the QCA Designer tool.
- The proposed circuit is designed with QCA technology using QCADesigner 2.0.3 version software with efficient clocking schemes.
- Simulations are generated for proposed design to check its design, implementation and functionality.
- The area and delay are calculated using QCADesigner software.

The rest of the article as follows section: 2 describes Terminology, section: 3 shows proposed work, section: 4 reviews comparison and section: 5 concludes the paper.

2. Terminology

QCA cell serves as an integral component in the process of QCA circuit designing [7]. Its structure is equivalent to a square box. Four corners of which contain four quantum dots. The location of electrons determines the two basic binary states possessed by the cells which are logic 0 and logic 1. There is no charge transport mechanism between the QCA cells, unlike CMOS. QCA comprises coulombic processes only [8]. The state of the nearby QCA cells during previous clock cycle determines the instantaneous state of a cell at a specified time [9]. Clocks numbers are the four clock zones present in QCA. The colour green is used to indicate clock zone 0, which is termed as the switch state [10]. The colour magenta is used to indicate clock zone 1, which is termed as the hold state. The colour blue is used to indicate clock zone 2, which is termed as the release state.

2.1 Basic QCA Elements

The working principle of QCA cell is based on Coulombs law. Due to Coulomb force, two electrons interact with each other called as kink energy with the four quantum dots of a QCA cell using the help of the clock signal through the junction of tunnels. The two electrons are occupied in such a way that they are diagonal to each other due to interaction between two electrons [11]. The way the electrons are placed give the polarization of the cell. The polarization of a QCA cell can only be done in two ways. The polarization of QCA cell are -1 and +1. There are four preferred clocking schemes or zones with four different clock signals which help with the information flow and switching of QCA cells [12]. They are discussed as Clocking Zone 0, zone 1, zone 2, zone 3. In each clocking zone, there are four clock phases as Switch, Hold, Release and Relax. The four clocks are each in 90 degree phase difference with each other, which is shown in Fig. 1. One of four clock signal use for reference while other clocks are delays namely

1. Reference clock (0 phase)
2. Delay 1 clock (phase)
3. Delay 2 clocks (phase)
4. Delay 3 clocks (phase)

When a cell holds no binary information, the cell is electro-statically switched from a void or null state, clocking is achieved depending on its phase [13]. In switch phase, cell state is influenced by its neighboring cells. In lock phase, cell sate is not influenced by its neighboring cells, it's independent [14]. At the time of transfer of information cells in each clock zone pass through all the four phases present in that particular clock zone. After passing all four phases information transfer takes place.

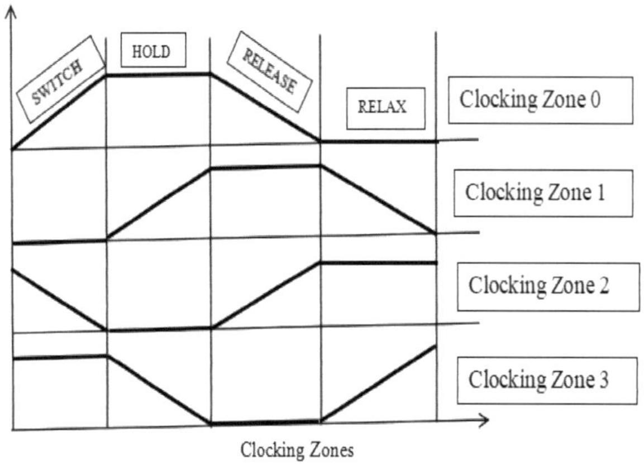

Fig. 1. Clocking Zones of QCA

Switch Phase
- During the switch phase, the junction barrier rise QCA cells are polarized respective to its polarization of the driver cell.
- Computation actually takes place in the switch clock phase

Hold Phase
- During the hold phase, the junction barriers are raised to its highest value.
- QCA cell is now input to the next stage in the hold phase

Release Phase
- During the release phase, the junction barriers are down or getting low.
- QCA cell is non-polarized and remain in the neutral state.

Relax Phase
- During the relax phase, the cell will return to its first reference clock phase.
- QCA cells can be again polarized in relaxed phase.

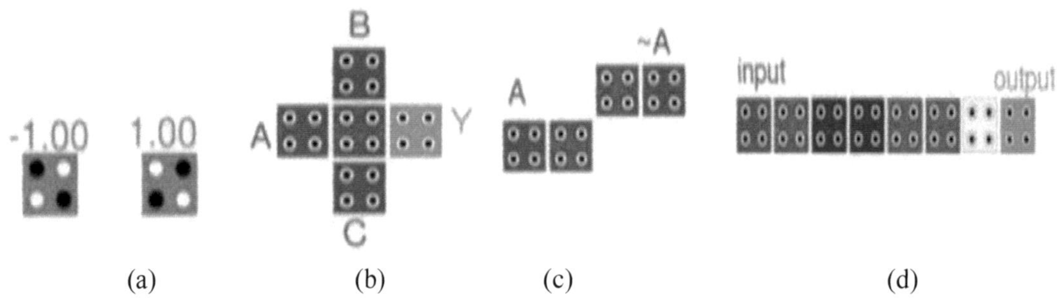

Fig. 2. QCA architectures based on quantum cell (a) Polarization (b) Majority gate (c) NOT (d) Wire

If we use all four clock signals in a circuit, then there will be a delay in the output of the circuit. So complex circuits have high delay. Assignment of clock schemes or zones is vital part during evaluation of QCA circuit and its simulation [17]. So, research of novel advanced clocking scheme is required during the implementation of QCA circuit layouts [18]. By arranging a different number of

QCA cells, different types of QCA devices can be designed such as polarization in QCA, QCA majority gate, QCA inverter, QCA wire, which is shown in Fig. 2a, 2b, 2c, 2d respectively. The data flow from one cell to another cell in a QCA wire. The array of QCA cells will act as a wire. If QCA cells are located adjacent to each other, the electron position is transferred from one cell to another adjacent cell. Due to this the state of all adjacent cells will all be same state because of coulomb interaction.

3. Proposed Work

3.1 Reversible Feynman gate.

In Feynman gate, if the first input is reset when the second output is same as input and if the first input is reset then second input is flipped at the output. It is also known as controlled not. Schematic of Feynman gate is drawn in Fig. 3. Cell layout and simulation results of Feynman gate are presented in Fig. 4 (a & b), respectively.

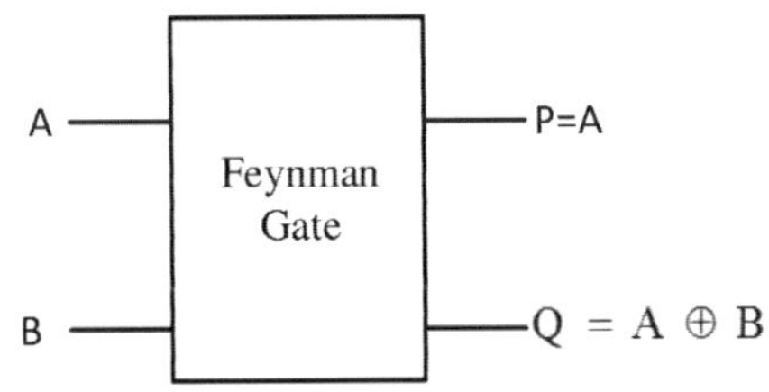

Fig. 3. Schematic of Feynman gate

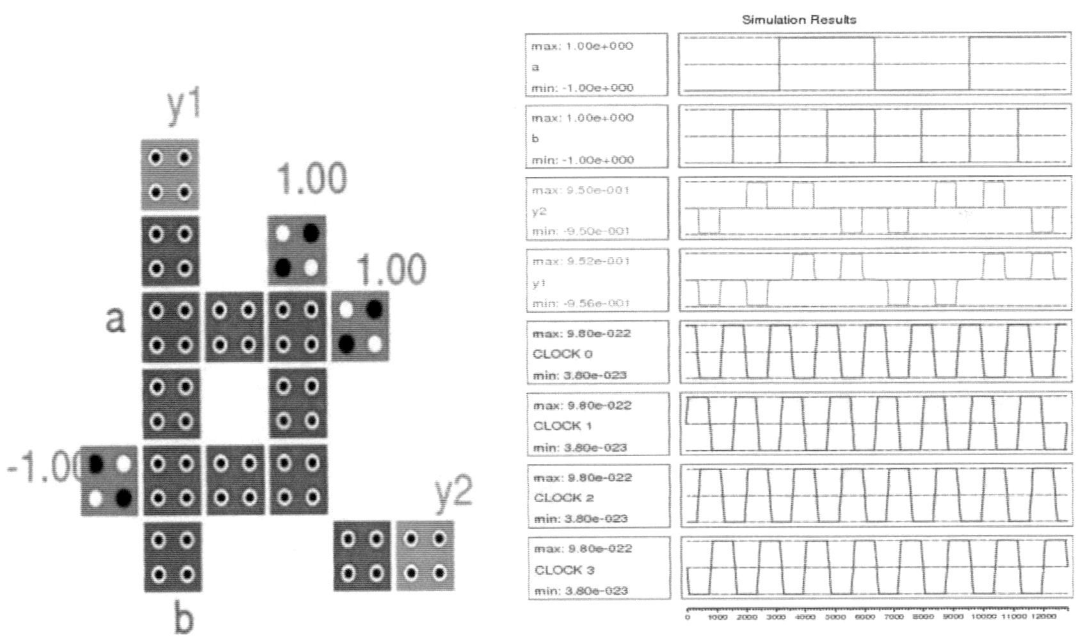

(a) (b)

Fig. 4. Feynman Gate (a) Cell Layout (b) Results

IOP Conf. Series: Materials Science and Engineering 1119 (2021) 012004 doi:10.1088/1757-899X/1119/1/012004

3.2 Reversible Fredkin gate

In Fredkin gate first output is same as the input. If the first input is set, then it swaps other inputs at the output. It is also known as a controlled permutation gate. In Fig. 5. shows a block diagram of Fredkin gate. Cell layout and simulation results of Fredking gate are presented in Fig. 6a and 6b respectively.

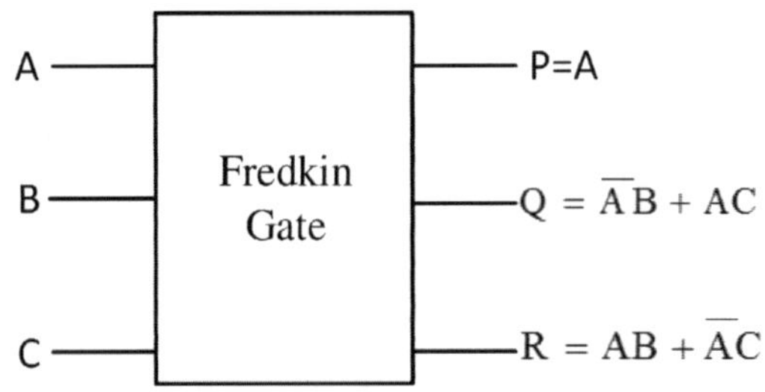

Fig. 5. Schematic of Fredkin gate

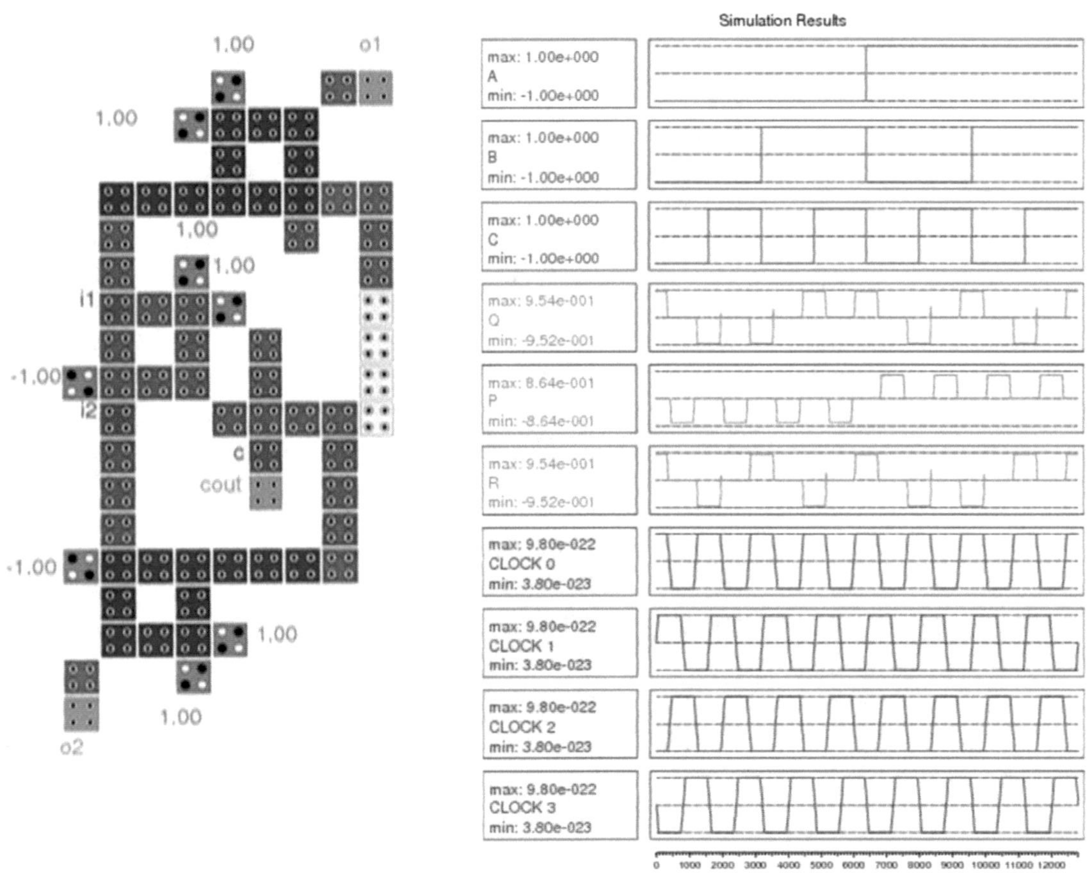

Fig. 6. Fredkin Gate Cell Layout and Results

3.3 Reversible Toffoli gate. In Toffoli gate, if the first two bits are set, then it flips the third input at the output. It is considered as a universal reversible logic gate and it is also known as a controlled-controlled-not gate. In Fig. 7. shows a block diagram of Toffoli gate. Cell layout and simulation results of Toffoli gate are presented in Fig. 8a and 8b respectively.

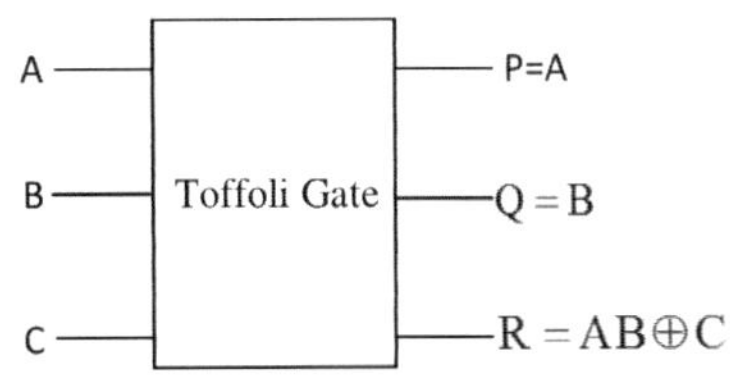

Fig. 7. Schematic of Toffoli Gate

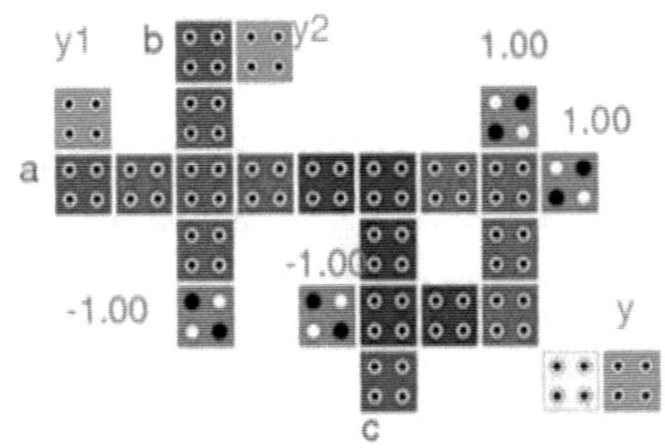

(a)

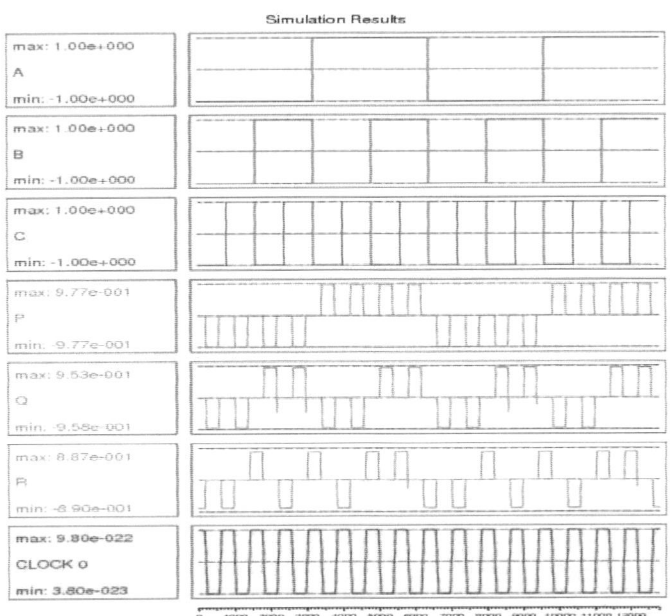

(b)

Fig. 8. Toffoli Gate (a) Cell Layout (**b**) Results

4. Comparison

The QCA integrating circuits are designed using basic logic gates like majority gates, inverters, and minority gates. These are important components, while designing any circuit in QCA technology. The design of architecture of computer system and processor system face problems regarding storage devices optimizing. For complex systems, efficient circuits with novel clock schemes need to be introduced to QCA technology designing. The designing of reversible gates in QCA technology is done in this article. The performance of proposed reversible gates based on QCA technology is evaluated and simulated using QCADesigner version 2.03 software and verified its functionality. Compared to prior works, the performance of the proposed reversible gate in QCA technology is improved in terms of cell, cycle and bounded area. The Table 1 shows the efficient utilization of the number of QCA cells by the proposed reversible Feynman, Toffoli, Fredkin gates than the previous best designs in the paper [15], [16], [19]. The novelty should be present in the architecture of basic reversible gates layout, because designing cells architecture using QCA technology is less complex, and high speed and low power during implementation. So, an effective circuit should be designed using basic components like inverter, majority gates with new techniques, which can make our circuit layout yield low power consumption even with high speeds. Designing in QCA technology requires advanced clocking and efficient circuit layout which can be tricky. The proposed basic reversible gates have advanced clocking, while designing circuits and use less number of QCA cells while keeping power low. The proposed basic reversible gate architecture in QCA technology should also be more efficient compared to existing designs. Then basic reversible gates designed using QCA techniques are compared to conclude which reversible gate architecture cell designed with QCA technology is more efficient.

Table. 1 Comparison results based on the QCA figure of merits

	Reference	**Feynman**	**Toffoli**	**Fredkin**
Cells	[15]	53	57	97
	[16]	89	128	*NS
	[19]	32	45	73
	Proposed work	17	25	67
Cycles	[15]	0.75	0.75	*NS
	[16]	1	1	*NS
	[19]	0.75	0.75	0.75
	Proposed work	0.5	0.5	0.75
Bounded-area	[15]	15*10	17*9	16*16
	[16]	17*15	31*16	*NS
	[19]	9*8	11*10	17*9
	Proposed work	6*6	10*6	18*8

*NS significance- not specified

5. Conclusion

The QCA technology is one of the popular Nanotechnology, which uses quantum gates for designing instead of transistors, which makes it a transistor less technology. Its attractive features like low density, low power consumption, high computation speed and high scaling over CMOS technology make it a future trend technology, which can used for designing electronic devices in the dimensions of nano-scale. The important concern during designing a basic reversible gates is to design an efficient circuit layout with advanced clocking schemes as its first concern for designing any circuit layout in the QCA technology. The basic reversible gates, which is of single layer architecture is proposed in this article. The proposed gate is designed and implement using QCA technology. The proposed three gates use 46.87%, 44.44% and 8.95% less number of QCA cells than the previous best design respectively. The proposed designs in this paper lesser rectangular layout of the gates than those earlier papers. Only one XOR gate used in Feynman gate while there is no usage of any majority gate. One from each of the XOR and majority gate are used in Toffoli gate and one majority gate and three XOR gates are used in Fredkin gate.

References

[1] Misra, N. K., Sen, B., Wairya, S., & Bhoi, B. (2017). Testable novel parity-preserving reversible gate and low-cost quantum decoder design in 1D molecular-QCA. *Journal of Circuits, Systems and Computers*, *26*(09), 1750145.

[2] Misra, N. K., Wairya, S., & Singh, V. K. (2014). Preternatural low-power reversible decoder design in 90 nm technology node. *International Journal of Scientific & Engineering Research*, *5*(6), 969-978.

[3] Landauer, R. (2000). Irreversibility and heat generation in the computing process. *IBM Journal of Research and Development*, *44*(1/2), 261.

[4] Bennett, C. H. (1973). Logical reversibility of computation. *IBM journal of Research and Development*, *17*(6), 525-532.

[5] Misra, N. K., Wairya, S., & Singh, V. K. (2014, May). An inventive design of 4* 4 bit reversible NS gate. IEEE International Conference on Recent Advances and Innovations in Engineering (ICRAIE-2014), 1-6.

[6] Bhoi, B. K., Misra, N. K., & Pradhan, M. (2018). Novel robust design for reversible code converters and binary incrementer with quantum-dot cellular automata. In Intelligent Computing and Information and Communication (pp. 195-205). Springer, Singapore.

[7] Lent, C. S., & Tougaw, P. D. (1993). Lines of interacting quantum-dot cells: A binary wire. *Journal of applied Physics*, *74*(10), 6227-6233.

[8] Misra, N. K., Wairya, S., & Singh, V. K. (2015). Frame of Reversible BCD Adder and Carry Skip BCD Adder and Optimization Using New Reversible Logic Gates for Quantum-Dot Cellular Automata. Australian Journal of Basic and Applied Sciences, 9(31), 286-298.

[9] Bhoi, B., Misra, N. K., & Pradhan, M. (2017). Design and evaluation of an efficient parity-preserving reversible QCA gate with online testability. Cogent Engineering, 4(1), 1416888.

[10] Misra, N. K., Wairya, S., & Sen, B. (2018). Design of conservative, reversible sequential logic for cost efficient emerging nano circuits with enhanced testability. Ain Shams Engineering Journal, 9(4), 2027-2037.

[11] Fredkin, E., & Toffoli, T. (2002). Conservative logic In: Collision-based computing, pp. 47-81.

[12] Dueck, G. W., & Maslov, D. (2003, March). Reversible function synthesis with minimum garbage outputs. In *6th International Symposium on Representations and Methodology of Future Computing Technologies* (pp. 154-161).

[13] Bahar, A. N., Habib, M. A., & Biswas, N. K. (2013). A novel presentation of toffoli gate in quantum-dot cellular automata (QCA). *International Journal of Computer Applications*, *82*(10).

[14] Islam, M. S., & Bahar, A. N. A Review on Reversible Logic Gates and it's QCA Implementation. *International Journal of Computer Applications*, *975*, 8887.

[15] Mohammadi, Z., & Mohammadi, M. (2014). Implementing a one-bit reversible full adder using quantum-dot cellular automata. *Quantum information processing*, *13*(9), 2127-2147.

[16] Garg, U., & Jain, R. (2016). Design and performance analysis of reversible RSG gate using QCA. *Int J Comput Appl*, *139*(12), 37-41.

[17] Walus, K., Dysart, T. J., Jullien, G. A., & Budiman, R. A. (2004). QCADesigner: A rapid design and simulation tool for quantum-dot cellular automata. *IEEE transactions on nanotechnology*, *3*(1), 26-31.

[18] Srivastava, S., Asthana, A., Bhanja, S., & Sarkar, S. (2011, May). QCAPro-an error-power estimation tool for QCA circuit design. In *2011 IEEE international symposium of circuits and systems (ISCAS)* (pp. 2377-2380). IEEE.

[19] Singh, G., Sarin, R. K., & Raj, B. (2017). Design and analysis of area efficient QCA based reversible logic gates. *Microprocessors and Microsystems*, *52*, 59-68.

Quality Parameter Index Estimation for Compressive Sensing Based Sparse Audio Signal Reconstruction

Vivek Upadhyaya[1], Girraj Sharma[2], Ashish Kumar[3], Sandeep Vyas[4], Mohammad Salim[5]

[1]Research Scholar, Malaviya National Institute of Technology, Jaipur
[2,3] Associate Professor, JECRC Foundation, Jaipur
[4] Professor, JECRC Foundation, Jaipur
[5]Professor, Department of ECE, Malaviya National Institute of Technology, Jaipur

2018rec9028@mnit.ac.in

Abstract. As we, all know that the size of data is increasing tremendously day by day. In a recent project, several petabytes were used to save an image of the Black Hole. Therefore, it is very crucial to develop a method that can reduce the size of data for transmission & storage purposes. The Traditional method for data compression & reconstruction requires so much data space, due to this problem another technique is proposed for the compression and recovery purpose. This method is termed Compressive Sensing (CS). As per the Nyquist sampling theorem, for proper reconstruction of the signal, we have to do sampling at the double rate of maximum data rate available in the signal. As a result, the storage requirement increased as well as the cost of the system was also enhanced. While on the other hand in Compressive Sensing, little samples are required for the reconstruction of the signal. So here in this paper, we have considered three music signals which are single tone, instrumental and vocal song. Values of Mean Square Error, Root Mean Square Error and Signal to Noise Ratio for different compression ratios mentioned in the tables and plots. By analyzing these values we can easily investigate the effectiveness of compressive sensing.

Keywords: Compressive Sensing (CS); Signal to Noise Ratio (SNR); Mean Square Error (MSE); Root Mean Square Error (RMSE); Restricted Isometry Property (RIP); Basis Pursuit (BP); Compression for Music Signals.

1. Introduction

Some standards form like JPEG, JPEG2000, MPEG, WAV, and MP3 associated with sparsity and compressibility give benefits to the transform coding schemes. Transform coding is a Leveraging but compressive sensing is the new combined groundwork for signal acquisition and sensor design. CS permits to reduce the sample of the signal which we want to sparse and compressed representation and evaluation cost of the signal. The states of the Nyquist Shannon sampling theorem need a few sample measurements to perfectly recover a band-limited signal but in the second case if the signal is sparse that's known as a basis, we extremely deduced the measurement sample signal which we seriously needed for recovery. As the result, when the signal is sparse it can be better regarding performances. So the compressive sensing is not used as a firstly sampled the signal at a high rate and after that compressed the data. We would like to search for a path that compressed data should be sense directly with a low sampling rate. Emmanuel Candès, Justin Romberg, and Terence Tao, and David Donoho, Who work on the CS in their research work trying the large-scaled with sparse and compressible represented signal easily reconstructed from the small number of linear and nonadaptive measurement sets [1][2][3]. The most challenging task in the CS field is designing the measurement method. The design of these measurement schemes and their dimension implement the practical data module and acquisition.

Many traditional and practical crises go along with the principle of the Nyquist sampling theorem for reconstruction and acquisitions of the image, video, audio from high dimensional data. To remove the conventional limits of sampling theory, compressed sensing (CS) recently days used in many fields of applied mathematics, computer science, and electrical engineering. The purpose of this paper gives a summary of the basic theory of compressive sensing. After the discussion of the overall introduction of the CS, let we start again discursion of the sparsity and other different low dimensional signal model. After that many questions were generated regarding this scheme that how to recover the complete signal from the few low dimensional signal accurately and give a high reliability guaranteed signal for different types of sparse recovery algorithms. We make a final result with the discussion of some sparse recovery framework. In the following work of the paper, we will take action to the fundamental for move on to some interesting and exploring directions. In which some new sensing design techniques with more advanced recovery conclusions are combined to describe the analog and discrete-time signal structure.

The paper is organized in such a manner that in section I the introduction about the techniques mentioned, in section II extensive literature review is given. Section III & IV depicts the problem statement and the objective of the paper. When we proceed to section V various properties & fundamentals are related to compressive sensing given. In section VI, the result and analysis mentioned followed by a conclusion in the next section.

2. Literature Survey

As we know, that audio processing is a very vast area, we can consider music processing as a tributary of audio processing. Various techniques evolved for speech processing but can also be used for music processing and provide good efficiency for the music signals. Because this paper contains a music signal for compression & recovery so we have to understand the basic parameter of the music signal. We can enhance the quality of the reconstructed signal by keeping these parameters in mind.

As per the paper, the author wants to propose a method by which the music signal processing or we can say audio signal processing will also depict the characteristics of the music or audio signal. The characteristic which is of interest is like pitch, melody, and timbre of the music signal. Here the consideration of music signal is like a very small part of the speech signal processing, and the method which is proposed by the author can give proper results like for the speech signal processing. All the qualities and characteristics which pertain to the speech signal also pertain to the music signal too [4].

In the paper of Masoumeh Azghani, Mostafa Karini proposed a multi hypothesis reestablishment scheme for a video sequence. In which it has a not only simple encoder but also a less complicated decoder [5]. MH technique involves the sparsity constraint and Tikhonov regularization. The author introduced the latest iteration algorithm which is based on MH. This iteration gives the best result for the performance during reconstruction. The computation process of Elaticnet is faster than the Thikhonov.

Compressed video sensing is focused on the paper for compression of a video sequence. In the single Hypothesis method, every block of the frame is defined and predicts the valve with the last resolving frame by this method. This is known as block matching because the encoder finds out the block of a particular frame with the highest matching. It has a drawback that it implicit the transmission at the transmitter side of the encoder to send the block motion vector that introduced the complexity at the transmitter side. So following issue is removing by the use of the multi hypothesis technique. In which the motion estimation task transmits from the encoder to decoder to be easy.MH- MC method gives a good efficient recovery production regarding cost and complexity at the receiver side. Further, it has some CVS problems Thikhonov regularization is offered with high PSNR. This is a combination of SH MH reconstruction strategies. Elastient based MH-HC was offered whose performance is good as compared to Tikonov regularization reconstruction.

As per the paper written by Mads Græsbøll Christensen, they applied a compressive sensing approach to the audio & speech signals. They used compressive sensing for the sparse decomposition of the speech signal with the help of window-based complex sinusoidal functions [6]. As the results discussed by the author the compressive sensing approach is valid for the sparse decomposition for audio & speech signals. But the problem they found is that the speech and audio signal have non-stationary components and complex too when the variation in sparsity is taken into the picture.

Azad M. Madni discussed how the artificial thinking of the system can improve the compressive sensing process and can also reduce the problems that arise in the sparse reconstruction of a signal [7]. He also discussed the sparse reconstruction algorithm. He found that the sparse recovery problem is a problem based on reverse engineering and is also distributed in various fields of information technology, like remote sensing & social networking.

As per the paper written by Ping-Keng Jao, Chin-Chia Michael Yeh, [8] they proposed two modifications in the LASSO technique which is used for the assignment of different code words used in the compressive sensing technique. The two modifications which are proposed by them, the first one make use of the repetitive nature of the music signal and the second thing is to optimize the screening constraint for signal recovery. The experiments done by them show clearly that the runtime which is required for the 10,000 codewords reduces up to the level of runtime required for 1000 code words. The conclusion which was proposed by them is that the larger dictionary size can improve the value of mean average precision.

In this paper, three music signals are considered as input signals on which compressive sensing is applied. As through the theory of Compressive Sensing, there are various parameters on which the reconstruction is dependent. The various parameters are like basis and sensing matrices and the type of reconstruction algorithm, which is used. So to see the effects of basis and sensing matrices on the reconstruction of music signal various types of Basis and Sensing matrices are taken and the results are discussed in this paper. Two parameters one is the MSE & the other is the SNR (Signal to Noise Ratio) considered for the quality detection purpose of the music signal.

3. Problem Statement

It is planned that this ultimate objective may be achieved by studying the following concerns: As we know that various areas like medical, error correction, RADAR signal processing require a huge amount of memory to store the data, it is the major problem with all these applications. Therefore, we aim to find out the solution to this problem. From the literature survey, we found that traditional sampling theory (Nyquist Rate) not able to sort out this problem because we have to require more than twice samples for the proper reconstruction of the signal. To resolve this storage scarcity, we have to take a few samples from the whole sample space and then reconstruct the original sample by some means. We have gone through the various approaches given in the literature and find out that there is a solution that can be able to solve this problem and that solution is called Compressive Sensing (CS). Now the work, which we want to do, is to study various parameters of Compressive sensing and find out the significance of those parameters. Our proposed work is to enhance the level of compression and then find out a good reconstruction of the original signal. We want to find out the relationship between the different parameters, which are the basic need for compressive sensing. The difference between traditional data compression and compressive sensing can be easily understood using the below-given diagrams.

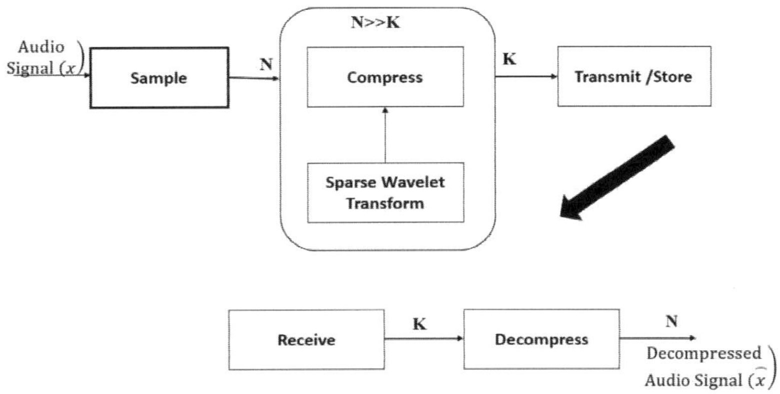

Figure 1: Sensing by Sampling

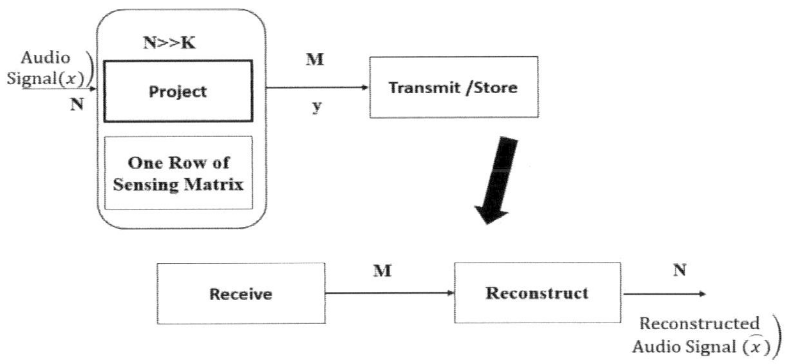

Figure 2: Compressive Sensing

4. Objective of Paper

The motive of this work is to provide a better approach to apply compressive sensing. Different parameter estimation approaches given in the literature. Our objective is to find out the essence of all these things and provide a better result to show the significance of this approach. As we know that the Basis matrix and Sensing matrix are two essential backbones of compressive sensing theory. One major thing is the reconstruction algorithm. Hence, to solve our purpose we have taken different sensing and measurement matrices and then applied reconstruction algorithms to them. Different signals are considered in which the very first signal, which is considered, is a Non-stationary single tone signal then we further move towards instrumental and vocal song signal. Then after enough iteration on the sound signal, we have gone through the change in basis & sensing matrix. The overall analysis of these signals provides different results for MSE, RMSE, SNR, etc because these are the quality parameters for any signal.

Throughout observation, we have certain conclusions and different statements, which show the significance of results. All significant quantities like SNR, MSE & RMSE are shown in tables for distinct random functions, and various plots associated with them are also shown in this paper. This whole analysis is very helpful to understand various results and the parameters shown in the results.

5. Reconstruction Properties

Some important properties related to compressive sensing stated below.

5.1. Restricted Isometry Property

If we consider it then the prime motive of encoding is to convert Nx1 K-sparse signal x to M x 1 measurement by incorporating an appropriate measurement matrix ϕ. The measurement matrix ϕ should project 2 distinct signals to 2 distinct sets of observations, so all the column sub-matrices of ϕ defined prominently [9].

Candès & Tao suggested a case for matrix ϕ which is known as the measurement matrix. For a K-sparse vector, a matrix kept the K-RIP if

$$(1-\delta_k)\|x\|^2 \leq \|\Phi x\|^2 \leq (1+\delta_k)\|x\|^2 \quad (1)$$

When the value of $\delta_K < 1$. Equation (1) signifies that all the submatrices of ϕ with K columns are well defined and show a good Isometry property. If $\delta_K <<$ 1measurement matrix ϕ has a massive probability to recover the (K/2) - sparse signal with a stable value of x. This condition is termed as Restricted Isometry Property (RIP) [9]. The relation between RIP & CS is well behaved if δ_{2K} is adequately less than 1. All pairwise distances between K-Sparse signals must be sustained in the measurement space.

$$(1-\delta_{2K}) \| x_1 - x_2 \|_2^2 \leq \| \Phi x_1 - \Phi x_2 \|_2^2 \leq (1+\delta_{2K}) \| x_1 - x_2 \|_2^2 \; \delta_{2K} \quad (2)$$

This implies that the above expression holds for all K-sparse vectors $\mathbf{x_1}$ and $\mathbf{x_2}$. Because $\mathbf{x_1}$ and $\mathbf{x_2}$ are, two distinct vectors and $\| x_1 - x_2 \|_2^2$ are invariably greater than zero. The value $\| \Phi x_1 - \Phi x_2 \|_2^2$ will never be zero. Therefore, to change the stability of sampling theory & get a K-sparse signal, the RIP constant δ_{2K} must be as small as possible. However, it is quite difficult to ensure that whether ϕ satisfies the equation (1) or not. As we go through the literature, we find out that luckily many random matrices fulfil the requirements of RIP condition the best suitable example is the Gaussian measurement matrix ϕ. An MxN i.i.d. Gaussian matrix will show RIP with great probability if $M \geq CK \log(N/K)$ where **C** is a constant [10-11] termed as RIP constant. The conclusion is that the N number of samples can be easily reconstructed using an M number of samples. This paper contains various basis & sensing matrices. The random matrices that are taken into consideration show nice RIP behavior if the signals have fair stability & high accuracy.

5.2. l_1 Minimization

As the discussion made in the last if the RIP holds, then l1 reconstructs K-sparse signals & does a decent calculation to assume the compressible signals with excellent probability using.

$M = O(L \log N/K)$ i.i.d. Gaussian measurements. Then the equation for recovery for l_0 norm will change to be:

$$x = \arg\min_{\tilde{x}} \| \tilde{x} \|_1, \text{ subjected to } \quad y = \Phi \tilde{x} \quad (3)$$

Eq. (3) is equivalent to the linear problem

$$\min \sum_{j=1}^{2N} v_j \text{, subjected to } v \geq 0,\; y = (\Phi, -\Phi)v \quad (4)$$

According to the theory, we can see that **v** has positive real coefficients & the size of **v** is 2N. The signal x is obtained from the solution **v*** of (4) via **x = (I,-I) v***. So there the conclusion is that l_1 minimization can easily be deduced using LPP. Other different methods like Interior-point, projected gradient [12] & iterative thresholding [13] can also be taken into consideration to deduce Eq. (4).

5.3. Greedy Pursuit

Greedy pursuit is the better option to recover the signal. It is a ceaseless signal reconstruction algorithm, to deduce the backing of the signal & it will provide an optimal choice of solution, which will fulfil the criterion, and it repeats until the condition is verified. The signal S recovered easily $x = (\Phi)^{\dagger} y$. Here Φ is the measurement matrix with the values of coefficients given by S & $(\Phi)^{\dagger}$ is the pseudo inverse of the measurement matrix Φ. If we define the pseud- inverse of a matrix Φ, which is a full rank in nature, then it is given $\Phi^{\dagger} = (\Phi^{*}\Phi)^{-1}\Phi^{*}$. Greedy Pursuit has very less response time while it is not so much stable & does not provide uniformity.

Compressive sensing is much favorable on an assumption that the l_1 minimization approach provides a better way for music signal compression & the way, which is utilized optimum in nature. Further work on this way is going on to find an algorithm, which is much faster & has very less response time with great recovery performance. Now here we provide a brief discussion about the recovery of signal x from the measured signal $y = \Phi x(+e)$ (here **e** denotes error) by using existing algorithms. Here we describe two types of algorithms first one is a variation on matching pursuit methodology and the others are Thresholding algorithms.

6. Result and Analysis

Here we apply the compressive sensing theory to the music signal. Basis Pursuit (BP) algorithm is used for recovery purposes. A MATLAB code is prepared for the application of the algorithm on the music signal. First of all, the signal is compressed using compressive sensing and then using the Minimum-l1 algorithm the reconstruction takes place. This basis pursuit algorithm has higher efficiency for the sparse recovery in the random domain. All three signals are considered in the form of a ".wav" file. The length of the music signal is 7 sec for all three domains i.e., in a single tone, instrumental & music. In the below section we are going to describe different values of MSE's for different signals and different measurement & basis matrices. Because due to this observation we can find out how much error will be there in the recovered signal concerning the value of the compression ratio (CR).

Table.1 CR v/s MSE values for Single-tone signal with different Basis and Sensing Matrices

MSE For Single Tone Signal								
CR (M/N)	**Gaussian**		**Poisson**		**Exponential**		**Rayleigh**	
	DCT	**DST**	**DCT**	**DST**	**DCT**	**DST**	**DCT**	**DST**
0.2	0.0033	0.0037	0.0034	0.0036	0.0034	0.0036	0.0035	0.0037
0.4	0.0007	0.0006	0.00079	0.00077	0.00079	0.00077	0.00076	0.00079
0.6	0.00021	0.00020	0.000249	0.00018	0.000249	0.00018	0.000241	0.00016
0.8	0.000049	0.000045	0.0000404	0.000045	0.0000404	0.000045	0.0000399	0.000040

Table.2 CR v/s RMSE values for Single-tone signal with different Basis and Sensing Matrices

RMSE For Single Tone Signal								
CR(M/N)	**Gaussian**		**Poisson**		**Exponential**		**Rayleigh**	
	DCT	**DST**	**DCT**	**DST**	**DCT**	**DST**	**DCT**	**DST**
0.2	0.0573	0.0605	0.0594	0.0607	0.0593	0.0603	0.0589	0.606
0.4	0.0276	0.0264	0.0279	0.0259	0.0281	0.0289	0.0248	0.0290
0.6	0.0147	0.0142	0.0157	0.0133	0.0158	0.0129	0.0149	0.0127
0.8	0.0065	0.0067	0.0064	0.0069	0.0064	0.0067	0.0062	0.0063

If we analyze tables 1, & 2 then we can see in most of the cases the value of MSE & RMSE is less in DCT than a similar case in DST. The question arises here that why there is a difference in the MSE & RMSE values for different measurement matrices. The answer to this question is very simple as we see that every random matrix has its PDF value. If we consider the Gaussian random matrix at a lower compression ratio then the value of MSE is lesser than other random matrices. According to the properties of random matrices, we can easily conclude that Gaussian provides the best results for compression. However, if we see the PDF curve of the Gaussian then we can analyze that when we increase the mean value then the PDF curve is much wider. That is why due to that the signal, which is to be applied here for transformation, interacts at a wider area so the MSE, which we get, has a lesser value than other random matrices. However, if we consider Poisson, Exponential & Rayleigh random matrices then for single tone signal the value of MSE is large in all these matrices because of their properties & PDF curve nature.

Table.3 Compression Ratio (CR) v/s MSE values for Instrumental Song signal with different Basis and Sensing Matrices

MSE For Instrumental Song Signal								
CR (M/N)	Gaussian		Poisson		Exponential		Rayleigh	
	DCT	DST	DCT	DST	DCT	DST	DCT	DST
0.2	0.0195	0.0201	0.0180	0.0191	0.0196	0.0198	0.0195	0.0197
0.4	0.0011	0.00135	0.00106	0.00119	0.00126	0.0013	0.0012	0.001256
0.6	0.0009	0.00099	0.00089	0.00095	0.00099	0.0010	0.00097	0.000985
0.8	0.000764	0.000812	0.00079	0.00084	0.000749	0.00083	0.00080	0.000821

Table.4 Compression Ratio (CR) v/s RMSE values for Instrumental Song signal with different Basis and Sensing Matrices

RMSE For Instrumental Song Signal								
CR(M/N)	Gaussian		Poisson		Exponential		Rayleigh	
	DCT	DST	DCT	DST	DCT	DST	DCT	DST
0.2	0.1396	0.1458	0.1341	0.1477	0.1452	0.1498	0.1364	0.1503
0.4	0.0761	0.0779	0.0698	0.0788	0.0742	0.0728	0.0726	0.0795
0.6	0.0389	0.0399	0.0345	0.0392	0.0323	0.0389	0.0312	0.0402
0.8	0.0277	0.0286	0.0281	0.0291	0.0274	0.0286	0.0283	0.0297

Now here we are analyzing the Instrumental song signal. The reason for the variation is the same as given in the previous section but here we can see that the values of MSE are much larger in comparison to single tone signal. The reason for this is that the instrumental song has higher frequency components than the single tone signal, and these all components are of different frequencies. So, these frequency components provide much more disturbance for each other. Therefore, reconstructing the original signal here is much more complicated. Therefore, it has larger error values than a single tone signal.

Table.5 Compression Ratio (CR) v/s MSE values for Vocal Song signal with different Basis and Sensing Matrices

MSE								
For Vocal Song Signal								
CR (M/N)	Gaussian		Poisson		Exponential		Rayleigh	
	DCT	DST	DCT	DST	DCT	DST	DCT	DST
0.2	0.0817	0.0896	0.0803	0.0835	0.0859	0.0896	0.0834	0.0865
0.4	0.0336	0.0357	0.0324	0.0341	0.0359	0.0384	0.0371	0.0389
0.6	0.0141	0.0153	0.0138	0.0143	0.0148	0.0151	0.0141	0.0147
0.8	0.0044	0.0046	0.0045	0.00465	0.00426	0.00436	0.0042	0.00441

Table.6 Compression Ratio (CR) v/s RMSE values for Vocal Song signal with different Basis and Sensing Matrices

RMSE								
For Vocal Song Signal								
CR (M/N)	Gaussian		Poisson		Exponential		Rayleigh	
	DCT	DST	DCT	DST	DCT	DST	DCT	DST
0.2	0.2859	0.3012	0.2834	0.2969	0.2931	0.2962	0.2888	0.2995
0.4	0.1833	0.1944	0.1824	0.1912	0.1869	0.1956	0.1865	0.1895
0.6	0.1187	0.1359	0.1126	0.1089	0.1029	0.1139	0.1069	0.1078
0.8	0.0663	0.0785	0.0661	0.0728	0.0648	0.0712	0.0651	0.0689

Here above in tables 5 & 6, we can easily observe that the values of Mean Square Error & Root Mean Square Error are larger than both single tone & instrumental song because the song signal has voice signal and other frequency components of different instruments. It will provide much more noise than the single tone & instrumental song signal. So here, it is too typical to reconstruct the original signal.

Different plots for Compression Ratio v/s SNR

Different plots that describe the relationship between the compression ratio and the SNR given below. For plotting these graphs different Basis and measurement matrices are taken into the consideration. The Basis matrices are DCT and DST, the sensing matrices which are taken are Gaussian, Poisson, exponential, and Rayleigh. We can see the difference between all the graphs significantly, which shows that SNR & CR are the parameters, which are directly proportional to each other.

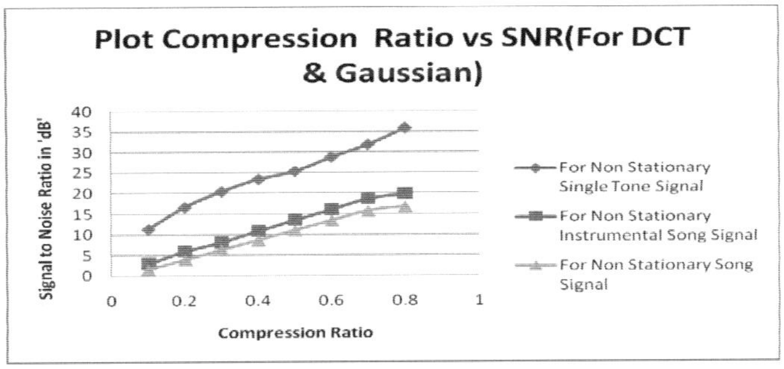

Figure.3 Plot for CR v/s SNR value For DCT as a basis & Gaussian as a sensing matrix for 3 Non-stationary music signals

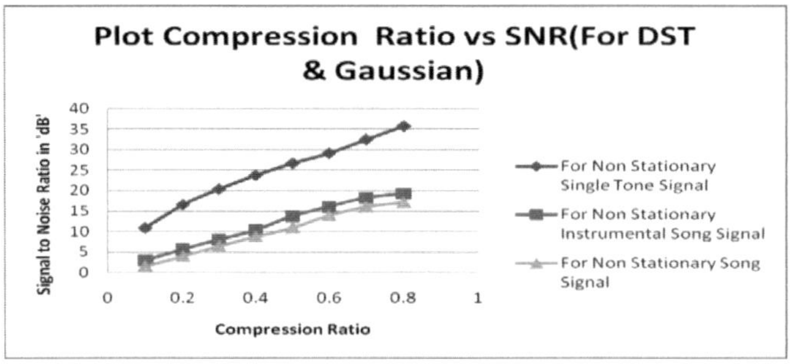

Figure.4 Plot for CR v/s SNR value For DST as a basis & Gaussian as a sensing matrix for 3 Non-stationary music signals

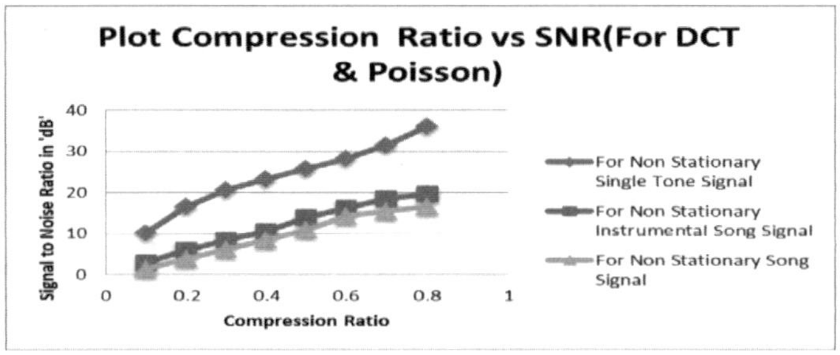

Figure.5 Plot for CR v/s SNR value For DCT as a basis & Poisson as a sensing matrix for 3 Non-stationary music signals

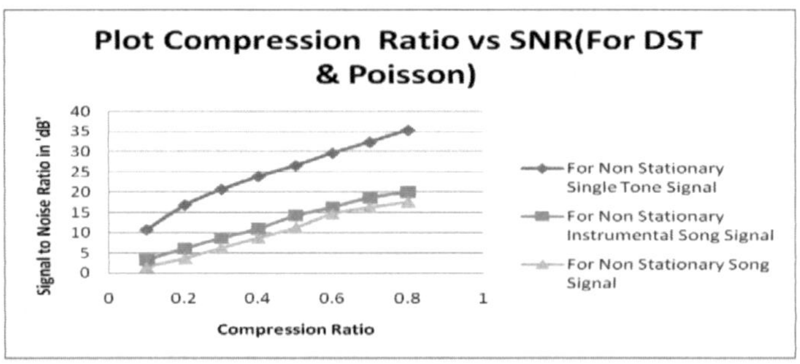

Figure.6 Plot for CR v/s SNR value For DST as a basis & Poisson as a sensing matrix for 3 Non-stationary music signals

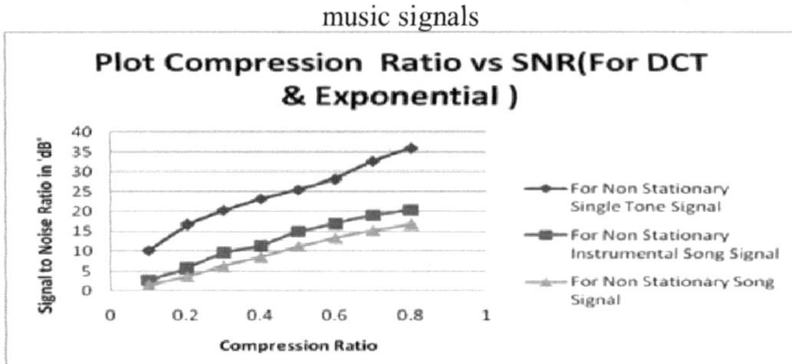

Figure.7 Plot for CR v/s SNR value For DCT as a basis & Exponential as a sensing matrix for 3 Non-stationary music signals

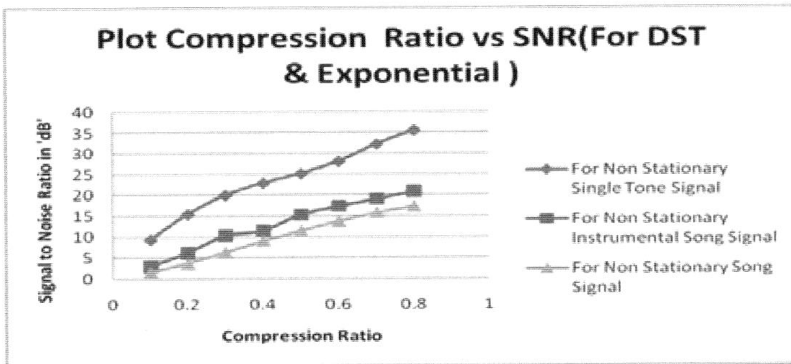

Figure.8 Plot for CR v/s SNR value For DST as a basis & Exponential as a sensing matrix for 3 Non-stationary music signals

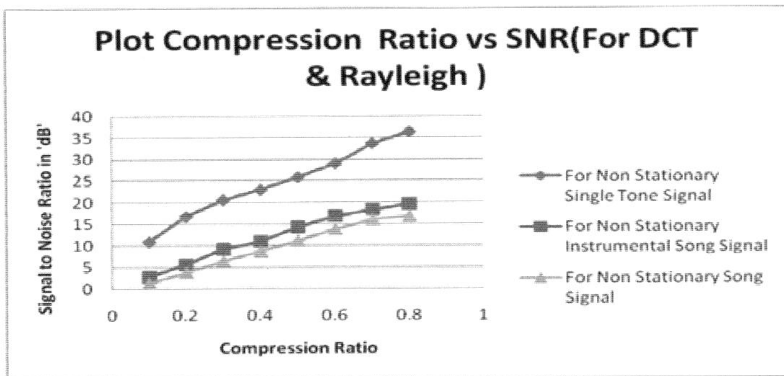

Figure.9 Plot for CR v/s SNR value For DCT as a basis & Rayleigh as a sensing matrix for 3 Non-stationary music signals

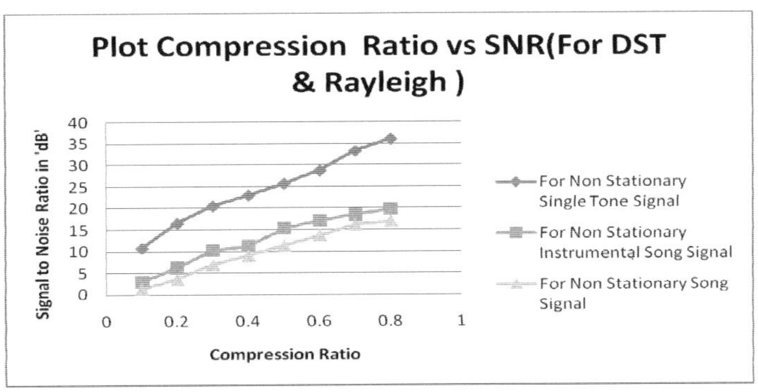

Figure.10 Plot for CR v/s SNR value For DST as a basis & Rayleigh as a sensing matrix for 3 Non-stationary music signals

7. Conclusion

It is a well-known fact that memory is a limited resource for any storage device. It is understood from the result and analysis section that we can recover the best quality of the actual music signal with very few samples. From the plots mentioned in the previous section, we can deduce that if we increase the compression ratio or we can say the number of measurements then the value of the signal to noise ratio also increases it means the quality of the reconstructed signal increases. From the result and analysis section, we can conclude that the value of the signal-to-noise ratio for the single tone signal is quite good than the instrumental and the vocal signal. It means that the complexity of the signal reduces the quality of reconstruction. Different parameters like MSE and RMSE are listed in Tables 1-6 in the previous section. From the tables, one point is very clear that with the variation in basis & sensing matrices & the type of signal values of SNR, MSE & RMSE also varies. The variations in the values of MSE & RMSE are due to the random distribution nature of Basis & Sensing matrices.

Acknowledgement: This research was "Supported by Visvesvaraya Ph.D. Scheme, MeitY, Govt. of India with unique awardee number “MEITY-PHD-2946” (Vivek Upadhyaya).

References

[1] Weiss, L., & McDonough, R. N. (1963). Prony's method, Z-transforms, and Padé approximation. *Siam Review*, *5*(2), 145-149.

[2] Klee, V. (1980). Another generalization of Carathéodory's theorem. *Archiv der Mathematik*, *34*(1), 560-562.

[3] Goluzina, E. G. E. (1985). Ranges of values of systems of functionals in certain classes of regular functions. *Mathematical Notes*, *37*(6), 438-442.

[4] Muller, Meinard, Daniel PW Ellis, Anssi Klapuri, and Gaël Richard. "Signal processing for music analysis." *IEEE Journal of Selected Topics in Signal Processing* 5, no. 6 (2011):1088-1110.

[5] Chen, Rui, Ying Tong, Jie Yang, and Minghu Wu. "Compressed Video Sensing with Multi-hypothesis Prediction." In *International Conference on Emerging Internetworking, Data & Web Technologies*, pp. 489-496. Springer, Cham, 2017.

[6] Christensen, Mads Græsbøll, Jan Østergaard, and Søren Holdt Jensen. "On compressed sensing and its application to speech and audio signals." In *Signals, Systems and Computers, 2009 Conference Record of the Forty-Third Asilomar Conference on*, pp. 356-360. IEEE, 2009.

[7] Madni, Azad M. "A systems perspective on compressed sensing and its use in reconstructing sparse networks." *IEEE Systems Journal* 8, no. 1 (2014): 23-27.

[8] Jao, Ping-Keng, Chin-Chia Michael Yeh, and Yi-Hsuan Yang. "Modified LASSO screening for audio word-based music classification using large-scale dictionary." In *Acoustics, Speech and Signal Processing (ICASSP), 2014 IEEE International Conference on*, pp. 5207-5211. IEEE, 2014.

[9] Joshi, Amit Mahesh, and Vivek Upadhyaya. "Analysis of compressive sensing for non stationary music signal." In *Advances in Computing, Communications and Informatics (ICACCI), 2016 International Conference on*, pp. 1172-1176. IEEE, 2016.

[10] V. Upadhyaya and M. Salim, "Basis & Sensing Matrix as key effecting Parameters for Compressive Sensing," 2018 International Conference on Advanced Computation and Telecommunication (ICACAT), Bhopal, India, 2018, pp. 1-6, doi: 10.1109/ICACAT.2018.8933745.

[11] Gao, Yun, Hao Zhou, and Xuejie Zhang. "Enhanced fast compressive tracking based on adaptive measurement matrix." *IET Computer Vision* 9, no. 6 (2015): 857-863.

[12] Zhao, Huihuang, Yaonan Wang, Xiaojiang Peng, and Zhijun Qiao. "Gradient- based compressive sensing for noise image and video reconstruction." *IET Communications* 9,no. 7 (2015): 940-946.

[13] Savic, Trifun, and Radoje Albijanic. “CS reconstruction of the speech and musical signal.’’ In Embedded Computing (MECO), 2015 4th Mediterranean Conference on ,pp.299-302. IEEE, 2015.

A Review on Land-use and Land-change with Machine Learning Algorithm

D Gupta[1]**, D Sethi**[2]**, R Bathija**[3]

[1,2]Department of Electronics and Communication, JECRC University, Jaipur, India
[3]Department of Electronics and Communication, JECRC College, Jaipur, India

[1]deveshgupta1603@gmail.com

Abstract. A significant proportion of the planet's ground environment has changed with ground use and land-related shifts, exacerbated by both human behavior and natural feedback. Anthropogenic activities have dramatically altered natural ecosystems, in particular in areas greatly affected by population growth and climate change such as Eastern Africa. For environmental protection and successful water management practices, it is important to be aware of the trends in land use & land cover (LULC). This study centered on developments in LULC patterns in Climate Modelling, remote sensing, and field data combined to identify positive feedback circuits or negative ones using remote sensing techniques and geographic information systems (GIS). Furthermore, its findings include statements that focus on policies that display the impacts and levels of LULC transformation and the dissemination of these improvements in time and space as a central element in the current methods for environmental control of changes and the management of natural resources.

Keywords: SVM, CNN, LULC, remote sensing, Spatial Transformation, Pixel-based change detection

1. Introduction

The transition from land use to land (LULC) has become a central and important feature of modern environmental change monitoring strategies and natural resource management [1]. Humans are causing climate changes to the Earth, which are resulting in fewer emissions for the biosphere. LULC and its programmers have been used to address human beings' physical, economic, cultural, and spiritual needs, and have experienced major improvements over the last 20 years. As part of the same, reductions in various vital resources such as water, soil, and plants that Western countries have undergone have contributed to decreased desirability and greater monoculture. Women, professionals, historians. Advanced natural resource research specialists use LULC experience to detect trends in natural capital, which requires natural capital changes in growth rates [2]. An awareness of land dynamics will help us gain a deeper understanding of the land. There have been several studies that have proved that improving the landscape level of land use and land cover (LULC) is enormously important for many purposes, including hydrology, agriculture, forestry, climate, geology, and ecology. Study on the shift of lumen diameters has also gained scientists' attention and momentum as well [3]. Most environmentalists conclude that the massive pollution of humans everywhere they go would lead to the imbalance of the atmosphere as well as to the consequences related to Climate Change (Global Warming).

2. Techniques for land use and land cover change

The use of geographical records, such as latitude and longitude, to map and identify areas of soil for which land cover and land use have been modified. Satellite-based Remote Sensing has revolutionized the study of land use and change over long periods because it has the potential of delivering synoptic information for land use and change over a given region at a given time. Being able to successfully explain how much soil is used or how much land is populated by various types of vegetation is helpful [5]. Geographical information systems allow me to construct a series of spatial fixes that correlate to the construction that drives the development. Remote sensing of environments can offer a lot of data on numerous physical features of the environment and can also provide data on whether changes in the landscape take place. It is a concept that defines the structure, distribution, and volume of changes, such as surface forms, boundary changes, and patterns at the time of transformation [6-9].

2.1 Digital change-based detection

2.1.1 Spatial Resolution

Due to its excellent time-frequency [10], computational digital format, synoptic vision, and a wider variety of spatial and spectral resolves, remote sensing data has been one of the key sources of shift detection studies. If we like 3 super spectral or SAR images or more, have the picture converted into greyscale images. When looking at spectral changes, a variety of factors, such as geographical, spectral, radiometric, time limits, atmospheric, and soil moisture areas may influence the detection of optical changes shown in Fig.1.The limiting factors are (a) geographical, spectral, radiometric resolution, (b) temporal limits, atmospheric conditions (c), and soil moisture conditions. [11]

Figure 1. Flow Diagram of Spatial Resolution

2.1.2 Spectral resolution

Although the shift-identification is a very important aspect of the identification of eruptions, more requirements need to be followed, such as visualizing multi-temporal images comprising high spatial and spectral resolution scopes; the requirement for accurate synchronizing of the radiometric and atmospheric conditions; and the set of the same spatial and spectral resolution images [13]. Crop diversity changes in the past can be calculated by the appearance or absence or showing the positions of the changes, the gradual trend of land cover changes throughout the long-term scheme, and the potential of adding precision by more changes in the landscape over time. Combining data sets from time-lapse images from several cameras, scientists were able to assess land use and discover trends for different species over time. Fig. 2 shows its flow process. Limitation: (a) there is not complete calibration of radiometric and atmospheric data, (b) the overlap between the data groups is troublesome, and, (c) it is very difficult to detect medium-scale vesicles to smaller vesicles [14].

Figure 2. Flow Diagram of Spectral Resolution

2.1.3 Radiometric change detection:

There are a variety of different approaches that can be taken to spectral analysis, from basic approaches such as band differentiation (a fundamental principle concerning the imaging capability of the image is an extensively sampled serial scan) to the advanced method (shown in Fig. 3) of using the vegetation indices [15]. He also defined the several slightly different ways in which variable sorting can be done to manage the aggregation of spectral groupings of THMs as a function of the scan acquisition date. The details of the distinctions between importance (detections) can be either a matter of simple binary changes (i.e., shift versus no change, such as image distinction, image rationing, etc.) or a thing of extensive changes (i.e., from time to time, as it is the case for comparisons after classification). This study aims to categorize change detection techniques into three categories based on image processing at pixel, function, and object-level by categorization scheme and are also classified by learning strategies in soft-computing approaches to detect change [16-18]. A Pixel is a numerical value for each band of an image or a simplified view of the contrast within that particular band of a particular image.

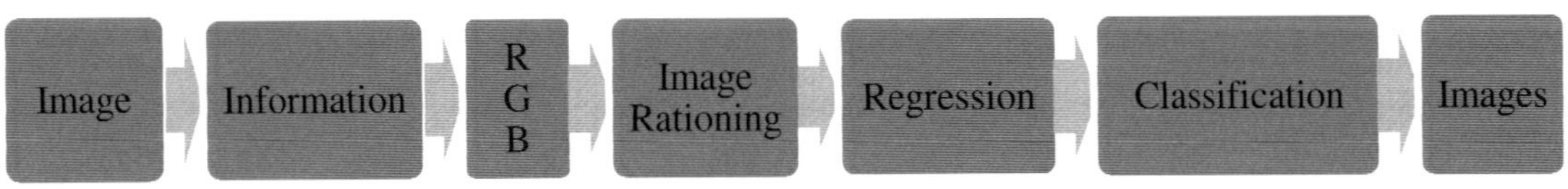

Figure 3. Flow Diagram of Radiometric change detection

2.2 Pixel level change detection techniques

Image pixels are a unit of measurement used in the image processing industry to define how much anything has moved by considering the spatial location of it. Different algorithms were built over time to render medical images in more detail. These are usually reduced in detail for lower resolution imaging. VHR image data is usually skewed due to different problems, (i.e., It is geo-referenced inaccurately, it has a wider variety of range of reflectance for each class, and it has very complex qualities when obtained – i.e., calibration issues) (e.g., sensing geometry, shading effect, and illumination angle). They used four different automatic image processing methods, most widely used to isolate post-classification and image ratio [16]. Deriving an image from a captured image is achieved by imaging by the pixel, then by erasing individual pixels bit by bit from the original image. In certain situations, a pixel can leap in one direction 18 times, and some cases can provide the condition where the pixel jumps up or down 17 times. Using traditional values of the newest image as a reference, a threshold value can be used to measure changes from the old to the current image, to determine how effective is the new-image recognition (see Fig. 4). To get the most precise estimate of how the threshold value corresponds to the halving of the exposure period, we took a close look at the change from pixel value to pixel value over time shown in Fig.4.These are the constraints on a) Errors in georeferencing, b) high difference in reflectance between grades, c) different acquisition features, and d) errors during the void inspection [17].

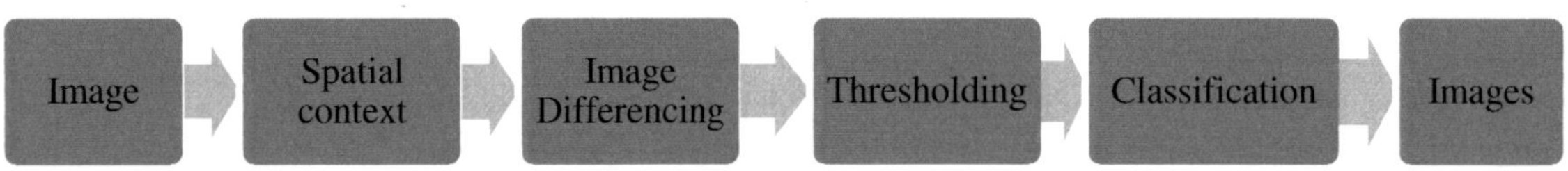

Figure 4. Pixel level change detection techniques

2.3 Image Regression

This method shows that a normal distribution exists between pixel values between 2.5 and 5 times in the same area. Because most of the pixels did not shift over the two years, the value is considered the same. The subject image (or its original value) is converted to the reference plate (the original image) through the least square regression through gain and offset (see Fig. 5). For the many other groups of images in the library, this transformation is performed similarly. It is a time-consuming method that delays the pace of classification [18].

Figure 5. Flow diagram Of Image regression

2.4 Artificial Neural Network

Neural Network is an algorithm that is non-parametrically supervised. It estimates the properties of data according to the training data shown in Fig. 6. The spectral data of the change time is the input used to train the neural network. A Back-propagation algorithm is also used to train the neural network model of multi-layer perceptrons. Neuro Network includes: (a) the hidden/back layer in ANN. Neuro Network is not well known; (b) the amount of training the data is important for network designing, and (c) Neuro Network functionality is not popular in image processing software [19-22]

Figure 6. Flow Diagram Artificial Neural Network

2.5 Support Vector Machine

The Data Distribution is not known to be [22] and the data sets are not displayed to avoid disclosing further information about the data. The machine learning algorithm is capable of learning small datasets for testing and is also able to do very high accuracy while handling larger data sets. The concept behind neural networks is that of interaction rules that are provided in memory links that are computed through dynamic associations between artificial neurons/elements. Neural networks can perform all kinds of tasks (shown in Fig. 7), but they have some drawbacks. The best use of neural networks is for text applications because their functionality can be translated to the text as well by the aforementioned [23].

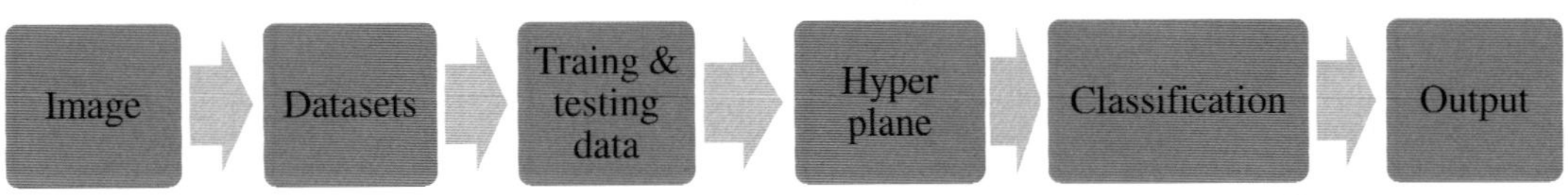

Figure 7. Flow Diagram Support Vector Machine

2.6 Decision Tree

Provided the classification of results, the decision tree algorithm is not parametric on data distribution or independence theory. Such algorithms create a structure of a flow chart in which every node tests several attribute values, every branch outputs a test result, and the tree leaves are classes or unit distributions. The classification of Decision tree-like nodes using the value of attributes that do the node classification is based on the value of the attributes. Assigning all data into classes, such as binary or multiclass, is mathematically [24] a need to do when making improvements to already categorized text. (see Fig.8)

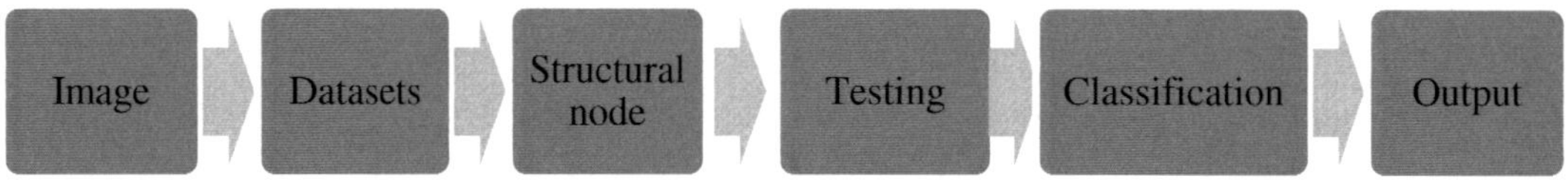

Figure 8. Flow Diagram Decision Tree

2.7 Fuzzy change detection

Smoothness is a mathematical methodology that uses the form of the curve to tell a tale about the variability between the classes and the measurable characteristics within the same class. Fuzzy reasoning results are not distinct and seamless; they are more likely and are confounded (see Fig. 9). The membership function takes elements closer to the membership function as probabilities rise. The P(E|B) of the new class is the higher the likelihood of the new class, the closer the elements are to the class. The class is mostly membered. It takes into account the similarities following in the classification of the results [25].

Figure 9. Flow Diagram Fuzzy change detection

2.8 Multi-sensor data fusion

For change detection, the multispectral image pixels are believed to be represented in multi-temporal spectral mixture analysis in terms of the sub-pixel proportions of pure spectral components that are related to surface constituents in a scene [26]. The technique addresses the heightened dimensionality problem that comes from higher spectral resolution. In a mixing model, synthetic particles can be mixed linearly with their constituent spectra weighted by the percent ground share. Using the Solans Vila & Barbosa Human Eye algorithm, it is important to have the accuracy of the measurements which come from a geometrically accurate location in the field for the dark-adapted visual response (end members), or in a laboratory for the spectral resolution. It can be extended to the increased dimensionality. A time-based calculation technique is used to resolve the dimensionality problem (see Fig. 10).

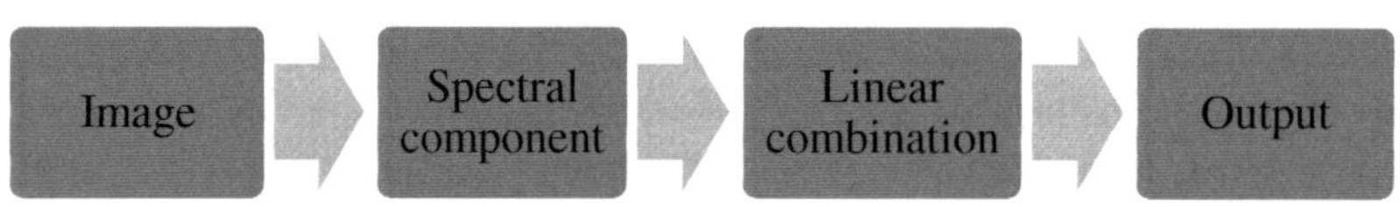

Figure 10. Flow Diagram Multi-sensor data fusion

3. Principal component analysis (PCA)

PCA is a way of working out what's in the data by converting the new data into a mathematically centered on the Principal Translation Axis, which avoids overlapping data. PCA generates an autocorrelated matrix by connecting covariable variables, equivalent to a correlation matrix that shows the correlation among variables [27]. The resulting matrix consists of the variables from decreasing values, and the main vectors are sorted as the majority of the variance in data is represented by the first variable. The key variables of the data set should then be uncorrelated variables describing the most important original information. The index is a three-dimensional device, made up of measures of luminosity, greenness, and humidity. It has a comprehensive (see detailed Fig. 11) map of the fields that are changing and of the thresholds used to define these changes.

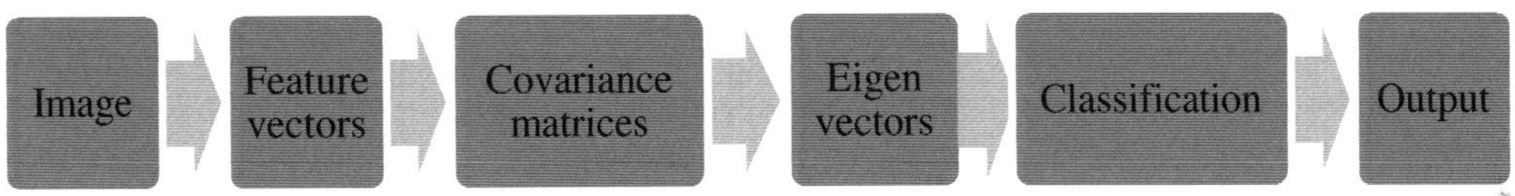

Figure 11. Flow Diagram Principal component analysis (PCA)

3.1 Change Vector Analysis

In the process of doing this, we have to first take multiple parallel slices, then a matrix of image samples will be generated. Before the revolution of the optical instruments, the conventional principle of vector analysis is based on the multi-temporal couples of spectral measurements, and the spectral transformation and appropriate magnitudes are the old parameters of a threshold. In spectral band metrics, pixel values are vectors instead of scalar values. The shift vectors are determined by subtracting pixel-sensitive vectors, much as in image differentiation. The map of the velocity vs. altitude was plotted to measure how the velocity shifted over time [28-30]. The two main reasons for vector shift analysis are (a) pixels are not as reliant on each other because the image sizes are predetermined, and (b) lots of pixels are required for the programmer to analyze. Any error in this technique can result in inaccuracy (see Fig 12).

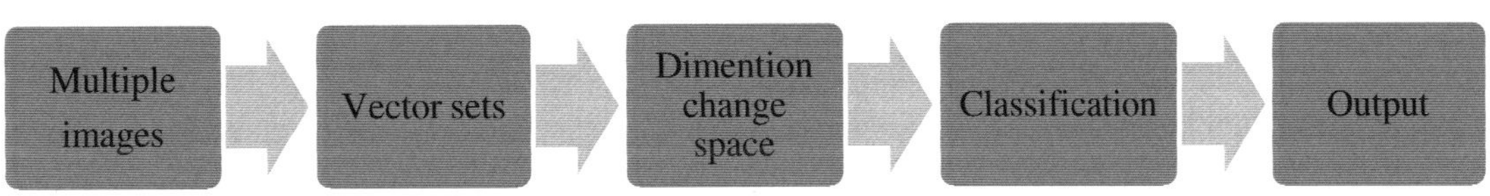

Figure 12. Flow Diagram Change Vector Analysis

3.2 Texture analysis-based change detection

This is a supercomputer-like technique that allows many image bands to be processed simultaneously. In the standard principle of vector analysis, the portion of the continuum that was measured is maintained by several time measurements, the size difference between a new and existent spectral sub-unit is the estimate of the threshold. In spectral band metrics, pixel values are vectors instead of scalar values. The shift vectors are determined by subtracting pixel-sensitive vectors, much as in image differentiation. The map of the velocity vs. altitude was plotted to measure how the velocity shifted over time. The two main reasons for transformation are calculated by matching the interpretations back with the textual meanings of the text. [30]

3.3 Object-Based Change Detection

The researchers [34-36] are concentrating on the description of images, class-artifacts, multi-temporal artifacts, and variations of the three. According to a study, replicated images were segmented with modifications based on the color of the object, its scale, and the other pertinent information. The study concluded that not only was the algorithm able to distinguish between different artifacts, it also effectively distinguished between two types of images: those with regular versus average color or texture or those that were blurred (e.g., image texture and geometry). The algorithm analyses a hash function to detect the rate at which meaningful items change from grey to grey. Connectivity analysis is used to remove objects from the master image, which are then compared to objects in a second image to see if there is a resemblance between the objects. To validate the same algorithm as the [31] algorithm, an approach called was used to define the relationship between two object-boundary pixels. It is said that the greatest benefit of this kind of segmentation is that because objects are derived by segmentation, algorithms are simple to implement and require direct object comparison to detect a shift. (see Fig. 13)

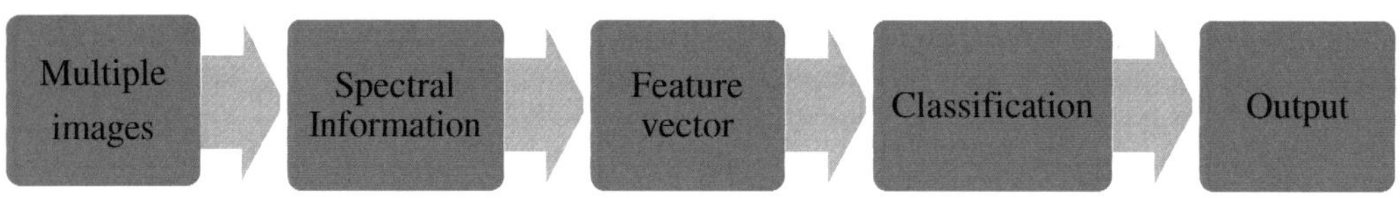

Figure 13. Flow Diagram Object-Based Change Detection

4. Hybrid Change Detection

Hybrid-shift recognition algorithms can make use of object recognition as well as pixel recognition. After the object matcher worked to return the object to the database, it then went on to find the best match (matched to the nearest object) through wider sets. This research didn't look solely at the variations between the two items. Instead, this study looked at both the Landsat Thematic Mapper (TM) and Landsat Improved Thematic Matter to measure the shift between both (ETM). He uses vector support systems and advanced methods to achieve an object-based classification. He contrasts the land use vector data and some objects to classify improvements. Correlation, grouping, particularity, and correlations jointly associated with each other through wavelets, all of which function together to represent a pattern, can all be represented as counting numbers, colors, and groups of letters. In the following, the area was segmented into artifacts [32] only by eliminating all of its surfaces when changes took place. Soft computing techniques for changing detection Soft computing is a mixture of modeling methodologies that make real-life solutions that are not mathematically modellable or too difficult to model. Soft computing, as it is called, uses techniques like pointless logic, evolutionary computation, and neural networks to achieve its goals.

Table 1 Analysis of Different Techniques

Author /Year	Worked on	Tool	Result/ Accuracy
Zhang et.al 2000	Unsupervised Classification	Unsupervised Classification	Classification of images, using remote sensing and geographic Information system tools with 88 Percentage.
Rundquist et.al 2001	Supervised classification	Supervised classification	Image classification, with the help of remote sensing and geographic information system software with 90 Percentage.
Lu et.al 2004.	Fuzzy classification	Fuzzy classification	Image classification, with the help of RS /GIS software with 84 Percentage.
Sudhira et.al 2004	The spatiotemporal analysis of LC dynamics	Spatial-temporal	Emphasis on population development/sprawl and rural land loss with 90 Percentage.
EEA 2006	Urban and rural growth fight for the same land	GIS-based	Cities expansion has typically taken place on former agricultural use with 91 Percentage.
Dewan et.al 2009	Multi-temporal RS datasets	Neural Network	Sustainable landscape Arti. Neural Networking and management. With 87 Percentage.
Huang et.al 2009	Several cities rise rapidly on their outskirts	Land stat	Transforming the surrounding rural areas into dense industrial and commercial ones with 89 Percentage.

4.1 Self-organising feature map network (SOFM)

This network has been planned to be very advanced with complex architecture. The job is to complete a small number of inputs and to assemble them into input patterns of relevant sizes, as in a neural network. This is done by the input neurons, which capture each mapped contact as an input, and the output neurons, which capture a pixel-by-pixel discrepancy. More comprehensive research will be conducted for the conversion of multi-sensor images and the use of multiple sensor data to use image recognition and pixel-based shift detection techniques [33] for the study of these modern multi-target capabilities. More advanced shift detection technology is required to identify minor shifts in textual elements in a frame. Some of these cameras can detect shifts in photo content, but there is not yet a proper mathematical model to integrate this information.

Table 2 Analysis of Different Techniques

Author /Year	Worked on	Tool	Result/ Accuracy
Murgante and Danese 2011.	Rising rural demand as a result of land leasing and the disappearance of some traditional roles outside the cities.	Pressure	With 87 Percentage.
Boori et.al 2014	Analyzed land-use related disruptions using a variety of RS and GIS techniques	GIS	With 89 Percentage.
Rawat and Kumar 2015	To map out the status of land use/cover of one of the development blocks	Segmentation	With 88 Percentage.
Turner et.al. 2001; Butt et.al 2015	Variation in land use/cover in a watershed can affect groundwater quality and source	Watershed	Higher surface runoff decreased groundwater recharge and application of chemicals by 90 Percentage.
Abdulhakim Mohamed Abdi 2017	Used Sentinel-2 Multispectral Instrument	SVM	Support vector machines and random forests are widely used in the remote sensing community for the use of mono-temporal vs. multitemporal imagery in land cover and land-use classification. with 91 Percentage.
Scott et.al 2019	Multiscale LU RS images and designed CNN structures	CNN	Multi-view CNN model achieved with 98.5 percentage.

4.2 Proposed Advanced computational data processing algorithms

Deep learning, smart agents, and human-computer interface (HCI) converge to provide unique goods. The study of the multiple processes in this field is a good way to explore natural processes and anthropogenic activities that promote land use. Neuroparalytic and the Cybernetics experiments. The region with the study area was close investigated, and over the past four decades, substantial data were analyzed using uncontrolled classification and satellite imagery, which shows to have set up a natural, unaltered landscape [34]. For certain land-use classification activities, neural networks (NNs) were increasingly used, but they are still under scrutiny by the science community for their supremacy. This research will aim to analyze the efficiency of CNN land classification approaches and to recognize changes in land use (LU). Eight CNN-based Transmitted Models for LU Scene Classification on Remote Sensing Data were extensively checked with three pretrained Alex Net, Google Net, and VGGNet CNN models [35-36].

5. Conclusion

With remote sensing images from various sensors acting as a basis, remote sensing-based measurement and mapping of the evolving design on LULCs in the Wami River basin taking prominence, and the same thing accomplishing contributions in both upstream and downstream transition. By observing the evolution of land distribution (and distribution of economic spheres) over time, it would be possible to define the general stages of the land distribution process. A comprehensive study on the transformation dynamics of military organizations and locations will provide advice to the military in its preparation, management, and security. However, the use of high-resolution multispectral satellite imagery will offer more information throughout the enhancement of the field. Although the images are not flawless, the device needs to use the method of minimizing the effects of a small flaw, by decreasing "salt and pepper" noise, by adding background correlations and information on the form of the product, by reducing the shading effects and geometric variations of the sensor viewing which will create better results than the pixel viewing. The recently created applications in this area provide new horizons for future computing. Back in the 1980s and 1990s, technical advances led to applications reaching beyond traditional boundaries, which then led to the growth of new companies, which in turn became organizations that are dependent upon remote sensing technology for further growth.

References

[1] Rawat, J. S., Biswas, V., & Kumar, M. (2013). Changes in land use/cover using geospatial techniques: A case study of Ramnagar town area, district Nainital, Uttarakhand, India. *The Egyptian Journal of Remote Sensing and Space Science*, *16*(1), 111-117.

[2] Kumari, M., Das, A., Sharma, R., & Saskia, S. (2014). Change detection analysis using multi-temporal satellite data of Poba reserve forest, Assam, and Arunachal Pradesh. *International Journal of Geomatics and Geosciences*, *4*(3), 517-525.

[3] Lambin, E. F., Turner, B. L., Geist, H. J., Agbola, S. B., Angelsen, A., Bruce, J. W., & Xu, J. (2001). The causes of land-use and land-cover change: moving beyond the myths. *Global environmental change*, *11*(4), 261-269.

[4] Birhane, E., Ashfare, H., Fenta, A. A., Hishe, H., Gebremedhin, M. A., & Solomon, N. (2019). Land use land cover changes along topographic gradients in Hugumburda national forest priority area, Northern Ethiopia. *Remote Sensing Applications: Society and Environment*, *13*, 61-68.

[5] Sankhala, S., & Singh, B. (2014). Evaluation of urban sprawl and land use land cover change using remote sensing and GIS techniques: a case study of Jaipur City, India. *International Journal of Emerging Technology and Advanced Engineering*, *4*(1), 66-72.

[6] Hegazy, I. R., & Kaloop, M. R. (2015). Monitoring urban growth and land-use change detection with GIS and remote sensing techniques in Daqahlia governorate Egypt. *International Journal of Sustainable Built Environment*, *4*(1), 117-124.

[7] Chomitz, K. M., & Kumari, K. (1998). The domestic benefits of tropical forests: a critical

review. *The World Bank Research Observer*, *13*(1), 13-35.

[8] Bruijnzeel, L. A. (2004). Hydrological functions of tropical forests: not seeing the soil for the trees?. *Agriculture, ecosystems & environment*, *104*(1), 185-228.

[9] Cheruto, M. C., Kauti, M. K., Kisangau, D. P., & Kariuki, P. C. (2016). Assessment of land use and land cover change using GIS and remote sensing techniques: A case study of Makueni County, Kenya.

[10] Foley, J. A., DeFries, R., Asner, G. P., Barford, C., Bonan, G., Carpenter, S. R., ... & Snyder, P. K. (2005). Global consequences of land use. *science*, *309*(5734), 570-574.

[11] Minale, A. S. (2013). Retrospective analysis of land cover and use dynamics in Gilgel Abbay Watershed by using GIS and remote sensing techniques, North-western Ethiopia. *International Journal of Geosciences*, *4*(07), 1003.

[12] Meesha, T. W., Tripathi, S. K., & Khari, D. (2016). Analyses of land use and land cover change dynamics using GIS and remote sensing during 1984 and 2015 in the Beressa Watershed Northern Central Highland of Ethiopia. Modelling Earth Systems and *Environment*, *2*(4), 1-12.

[13] Adeel, M. (2010). Methodology for identifying urban growth potential using land use and population data: A case study of Islamabad Zone IV. Procedia Environmental Sciences, 2, 32-41.

[14] Rawat, J. S., & Kumar, M. (2015). Monitoring land use/cover change using remote sensing and GIS techniques: A case study of Hawalbagh block, district Almora, Uttarakhand, India. *The Egyptian Journal of Remote Sensing and Space Science*, *18*(1), 77-84.

[15] Weng, Q. (2001). Remote sensing? GIS evaluation of urban expansion and its impact on surface temperature in the Zhujiang Delta, China. *International journal of remote sensing*, *22*(10), 1999-2014.

[16] Hassan, Z., Shabbir, R., Ahmad, S. S., Malik, A. H., Aziz, N., Butt, A., & Erum, S. (2016). Dynamics of land use and land cover change (LULCC) using geospatial techniques: a case study of Islamabad Pakistan. *SpringerPlus*, *5*(1), 812.

[17] Liang, S., Fang, H., Morisette, J. T., Chen, M., Shuey, C. J., Walthall, C. L., & Daughtry, C. S. (2002). Atmospheric correction of Landsat ETM+ land surface imagery. II. Validation and applications. *IEEE transactions on Geoscience and Remote Sensing*, *40*(12), 2736-2746.

[18] Ayele, G. T., Demessie, S. S., Mengistu, K. T., Tilahun, S. A., & Melesse, A. M. (2016). Multitemporal land use/land cover change detection for the Batena Watershed, Rift Valley Lakes Basin, Ethiopia. In *Landscape Dynamics, Soils and Hydrological Processes in Varied Climates* (pp. 51-72).

[19] Lambin, E. F. (2001). Global land-use and land-cover change: what have we learned so far?. *Global Change News*, *46*, 27-30.

[20] Bewket, W. (2002). Land cover dynamics since the 1950s in Chemoga watershed, Blue Nile basin, Ethiopia. *Mountain research and development*, *22*(3), 263-269.

[21] Gimblett, H. R. (2005, December). Modeling Human-Landscape Interactions in Spatially Complex Settings: Where are we and where are we going. In *MODSIM 2005 International Congress on Modelling and Simulation, Modelling and Simulation Society of Australia and New Zealand.*

[22] Alemayehu, F., Taha, N., Nyssen, J., Girma, A., Zenebe, A., Behailu, M., & Poesen, J. (2009). The impacts of watershed management on land use and land cover dynamics in Eastern Tigray (Ethiopia). *Resources, Conservation and Recycling*, *53*(4), 192-198.

[23] Tiwari, M. K., & Saxena, A. (2011). Change detection of land use/land cover pattern in and

around Mandideep and Obedullaganj area, using remote sensing and GIS. *International Journal of Technology and Engineering System*, *2*(3), 398-402.

[24] Bakr, N., Weindorf, D. C., Bahnassy, M. H., Marei, S. M., & El-Badawi, M. M. (2010). Monitoring land cover changes in a newly reclaimed area of Egypt using multi-temporal Landsat data. *Applied Geography*, *30*(4), 592-605.

[25] Abd El-Kawy, O. R., Rød, J. K., Ismail, H. A., & Suliman, A. S. (2011). Land use and land cover change detection in the western Nile delta of Egypt using remote sensing data. *Applied Geography*, *31*(2), 483-494.

[26] G Maseko FE Lindsay SN Gowan, J. (2000). Dynamics of urban growth in the Washington DC metropolitan area, 1973-1996, from Landsat observations. *International Journal of Remote Sensing*, *21*(18), 3473-3486.

[27] EEA (European Environment Agency) (2006) - Urban sprawl in Europe: The ignored challenge. EEA Report No. 10/2006, Copenhagen, Denmark.

[28] Abdi, A. M. (2020). Land cover and land use classification performance of machine learning algorithms in a boreal landscape using Sentinel-2 data. *GIScience & Remote Sensing*, *57*(1), 1-20.

[29] Dewan, M. Z., Galloway, A. E., Kawashima, N., Dewyngaert, J. K., Babb, J. S., Formenti, S. C., & Demaria, S. (2009). Fractionated but not single-dose radiotherapy induces an immune-mediated abscopal effect when combined with anti–CTLA-4 antibody. *Clinical Cancer Research*, *15*(17), 5379-5388.

[30] Fichera, C. R., Modica, G., & Pollino, M. (2012). Land Cover classification and change-detection analysis using multi-temporal remote sensed imagery and landscape metrics. *European journal of remote sensing*, *45*(1), 1-18.

[31] Sudhira, H.S., Ramachandra, T.V. and Jagadish, K.S. (2004) *Urban sprawl: metrics, dynamics and modeling using GIS*, International Journal of Applied Earth Observation and Geo-Information, 5(1):29–39.

[32] Huang, B., Zhang, L., & Wu, B. (2009). Spatiotemporal analysis of rural-urban land conversion. *International Journal of Geographical Information Science*, *23*(3), 379-398.

[33] Suwalka, I., & Agrawal, N. (2018). An improved unsupervised mapping technique using AMSOM for neurodegenerative disease detection. *International Journal of Computational Systems Engineering*, *4*(2-3), 185-194.

[34] Rawat, J. S., Kumar, M., & Pathak, R. N. (2013). Spatio-temporal dynamics of Almora town area, India. *Inter. J. Adv. Rem. Sens. GIS*, *2*(1), 425-432.

[35] Fichera, C. R., Modica, G., & Pollino, M. (2013). Integration of Satellite Remote Sensing Techniques and Landscape Metrics to Characterize Land Cover Change and Dynamics. In *Geographic Information Analysis for Sustainable Development and Economic Planning: New Technologies* (pp. 228-244). IGI Global.

[36] Yalew, S. G., Mul, M. L., Van Griensven, A., Teferi, E., Priess, J., Schweitzer, C., & van Der Zaag, P. (2016). Land-use change modelling in the Upper Blue Nile Basin. *Environments*, *3*(3), 21.

Application of SNORT and Wireshark in Network Traffic Analysis

G Jain[1], Anubha[2]
[1,2] Computer Science Department, Swami Vivekanand Subharti University, Meerut, UP, India
[1]jaingarima2011@gmail.com

Abstract. Now a days, our life is totally dependent on internet which in turn leads to different kind of intrusion. In order to detect intrusion, detection tool is needed so as to detect any kind of intrusion without user involvement. Intrusion detection system is a process of analyzing intrusion so as to provide secured data transmission in networking. One of the most used tools in analyzing packets is Wireshark. Nevertheless, Wireshark is less effective in detecting the intrusion still it can be used to intercept and detect network encrypted traffic which will analyze the captured traffic. Wireshark helps in identifying number of attacks like DoS, DDOS etc. SNORT take all live packets from internet and captured the live packets with predefined rules and if matches didn't find then directs alert messages to the user. In this method, router is used to access internet so as to get precise flow of data packets. It generates a log file which entails all the live captured packets. The log file "Wi-Fi" generated by SNORT tool is then exported to Wireshark to examine the captured network packets. The Wireshark tool creates each detail about the packets of the recorded log file. In this paper, the I/O graph outlined the flow of the packet which demonstrations the total traffic which is further measured in either bytes or packets per second. TCP errors and delta time graph are shown.

Keywords. SNORT, Wireshark, NMAP, data packets, network traffic, packet analysis, DoS, DDoS.

1. Introduction

Network protection is one in all the most important demanding situations that organizations are going through from time to time. There are plenty of tries via way of means of the black hat hackers to interrupt and compromise with the safety of Company's community and a number of them are even successful. As using net increasing, those malicious sports are gaining reputation many of the black hats. Everyday big quantity of records is being generated and exceeded on and plenty of those records holds touchy records approximately the organization and its employee. Thus, securing community is one of the maximum vital challenge for an organization to survive. To make this simpler and green we use Intrusion Detection System it allows to acquire records approximately any malicious packet that passes throughout an organization's community.

1.1. Intrusion Detection System

Intrusion detection system (IDS) is a sort of safety machine for computer systems and laptop networks. Intrusion Detection specifically enables in detecting outer and internal assaults done via way of means of both consumer and hackers. An IDS machine collects facts from numerous reasserts and analyzes facts from numerous regions inside a laptop or a community to be aware of viable safety breaches, which encompass each intrusion (assaults from outdoor the organization) and misuse (assaults from in the organization). IDS make use of vulnerability assessment (also known as scanning), to evaluate the safety of a laptop machine or community.

1.1.1 Prons of IDS
- Trace any changes made within the network.
- Audits system activity
- Normal and abnormal activities can be distinguished in the network

1.1.2 Cons of IDS.
- At times provides false alarms i.e., IDS might produce alert message even though packet is not malicious the packet wasn't malicious.
- Takes a lot time in time consuming
- Is not 100% harmless from attacks. at affiliations are as full and complete as possible and include the country.

1.2. SNORT
Snort as NIDS is an open-source network intrusion and sensing program (IDS / IPS). Snort is the mostly used as IDS / IPS technology in the world and brings collectively the benefits of protocols, signatures, and specifications based on exceptions. This tool is a IDPS (Intrusion Detection and Prevention System) which uses a sequence of rules/ policies that aid to outline or describe malicious network activity and make use of those rules/ policies to discover packets that didn't match and creates alerts for users. User has written these rules/ policies in a text file which is associated with snort.conf file where all the snort configurations are stated. To get SNORT running, few commands are these so that it can examine network behaviour.

1.3. WIRESHARK
The most widespread network protocol analyzer is Wireshark [2]. It has an efficient feature and runs on every platform comprising Windows, Linux, OS X, and UNIX. Network professionals, security experts, developers, and educators around the world use it regularly. It is an open source which is used by security experts, network professionals etc. on regular basis. Global team of protocol experts developed and maintained this tool. Wireshark previously used to be recognized as Ethereal. It is a free packet sniffer which has some of the aspects like analysis, network troubleshooting etc. This tool is used for viewing, capturing, and analysis of data packets. It has refined wireless protocol to aid administrators troubleshoot wireless networks. It can capture traffic from the air and decrypt it into that format which aids administrators to trace down those problems which causes poor performance, intermittent connectivity using appropriate driver support.

2. Literature Review
Author [1] executed a network intrusion system with the help of an anomaly tool by considering machine learning. This procedure logged the live network transmission using Wireshark and snort tool. Wireshark showed the I/O graph of network packets. The authors [2] presented a procedure which uses Wireshark to examine network protocols. The author has done examination to identify predictable network threats such as hidden FTP, port search etc. The authors also showed that packet inspection can discover an extensive type of protection vulnerabilities and attacks on networks. The authors [3] projected a framework to dodge DDOS attacks in cloud computing using Wireshark and Snort IDS/ IPS. Moreover, authors simplified the facilities available on-demand in cloud storage to permit the customer to access such facilities via internet. DOS and DDOS are the most substantial attacks which yield benefits of cloud computing vulnerabilities. The authors presented an experimental design to stop DDOS attacks but consistency and expense factors were not measured. The authors [4] presented the network protocol analysis to accumulate data for examination and technical understanding of packets. A complete study was presented about network attacks in terms of security, network sniffer and protocol

analyzer. The authors [5] planned technique to assessment the efficiency and accuracy of the DDOS attack. The paper observed important factors that impact the effectiveness and detection abilities of snort and planned schemes for refining the IDS. The research has proven that better tools can use to strengthen snort's handling effectiveness. Snort IDS are the attacking discovery device which the analysts around the globe intrigued. In [6-7] introduced the device that causes the organization head to make the SnortIDS rules and alarm by means of Graphical User Interfaces (GUI). The authors [8] clarified about Snort lab which assists the understudies with figuring out how to make the Snort-IDS rules. They gave 6 issues to the understudies by applying the Snort-IDS rules for each assaulting discovery. Likewise, In [9-12] authors examined subtleties of Snort-IDS rules to the planned and changed Snort-IDS rules for distinguishing and securing the assault in 3 sorts of site which are SQL infusion, XSS and order execution, individually.

3. Methodology

In the procedure, the final analyzed report of the captured live data packets is generated. The SNORT, an open-source tool will take live data packets from the internet. This tool will match the captured packets with predefined signatures and directs alert messages to the user if matches didn't find. In the procedure, in order to access internet, WIFI is used so as to get precise flow of data packets. After initiating the capturing process, this tool generates a log file in its directory that stores each and every captured packet [13-14]. This log file entails data packets along with their alert messages. The log file generated by SNORT tool is then exported to Wireshark to examine the captured network packets. The Wireshark tool creates each detail about the packets of the recorded log file. Given figure 1 describes the framework of Wireshark and snort. After then figure.2 explains the flow of methodology.

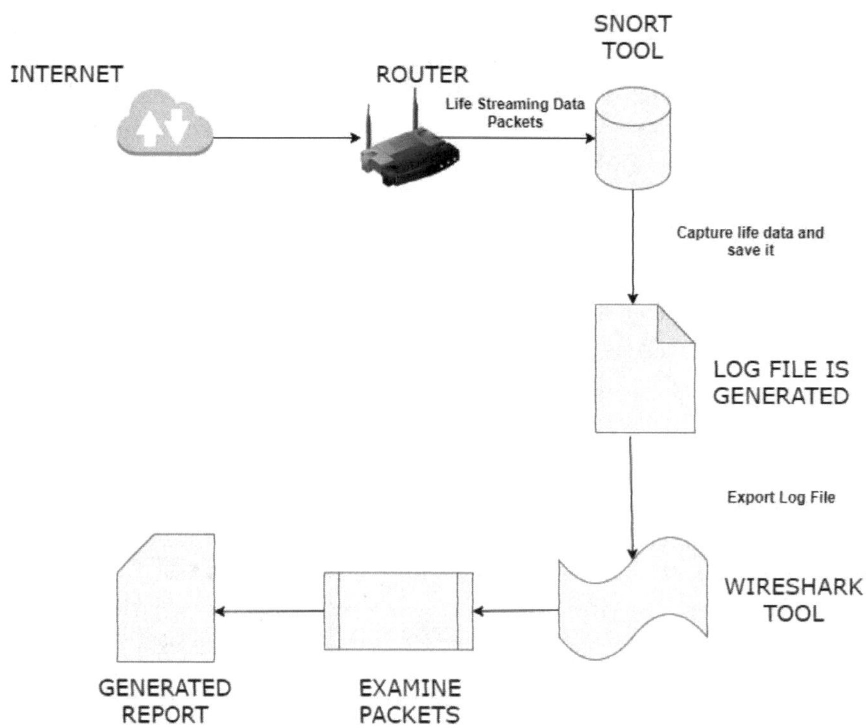

Figure 1. Framework of methodology using Snort and Wireshark

IOP Conf. Series: Materials Science and Engineering 1119 (2021) 012007 doi:10.1088/1757-899X/1119/1/012007

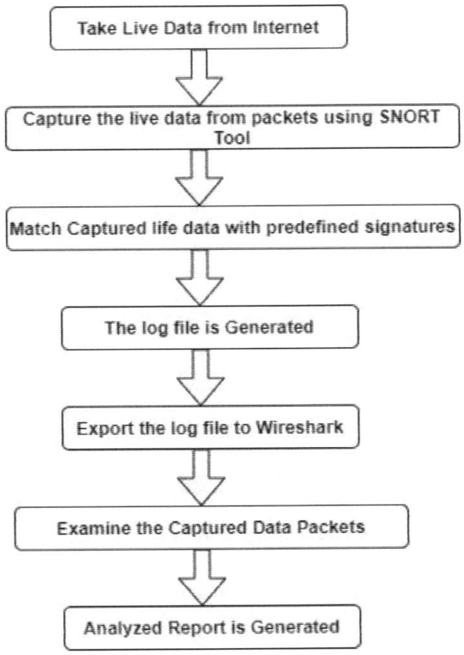

Figure 2. Flowchart of methodology

4. Experimental Results

To configure SNORT, firstly visit www.snort.org.com and also install wincap to use snort tool in windows. SNORT configuration file is divided into 9 sections which are as follows:

- Set the network variables.
- Configure the decoder.
- Configure the base detection engine.
- Configure dynamic loaded libraries.
- Configure preprocessors.
- Configure output plugins.
- Customize your rule set.
- Customize preprocessor and decoder rule set

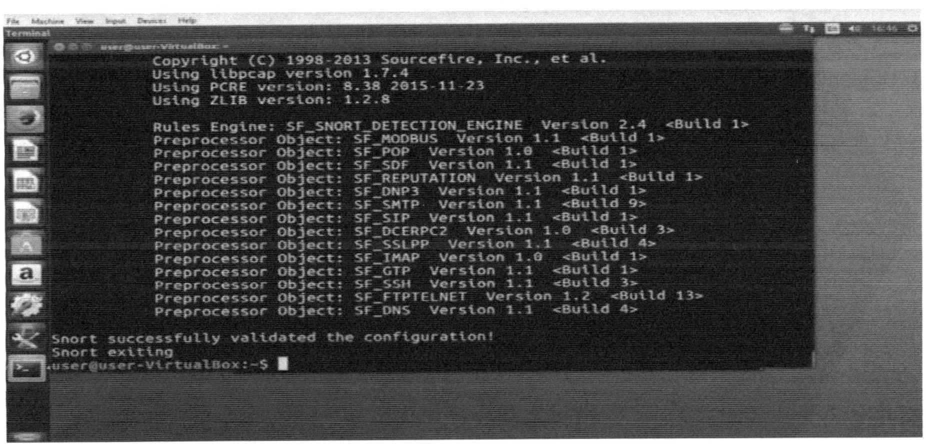

Figure 3. Successful SNORT configuration

In figure 3 Configuration using SNORT is done successfully with a successful message.

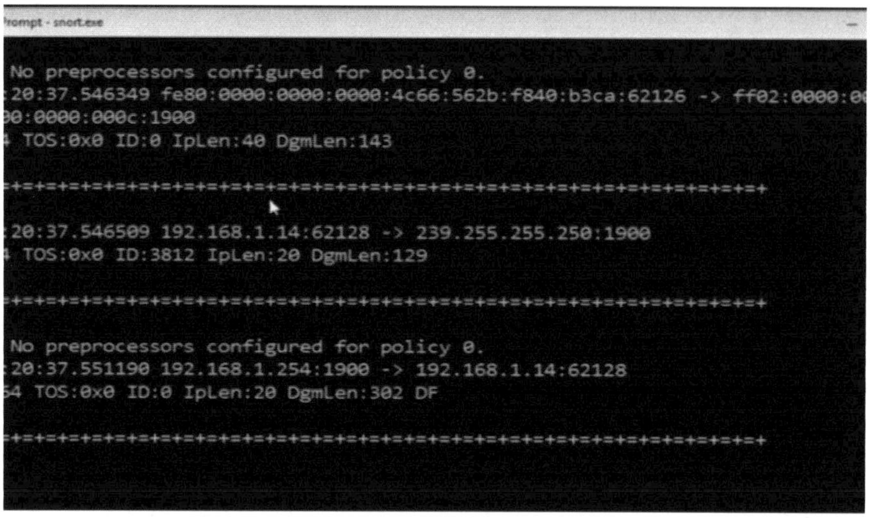

Figure 4. Live data packets captured by Snort with their alert messages

In figure 4 using snort data packets are captured with alert messages. By using "Ctrl+C", the process of capturing packet will be stopped and can have a summary.

Figure 5. Summary of captured data packets by Snort- NIDS

In figure 5 the summary of captured data packets is observed using SNORT-NIDS.

5. Export to Wireshark

The recorded log file is exported to Wireshark for an in-depth evaluation of the captured packet. In these different protocols are studied. Whenever the file is retrieved, all the collected data can be seen. It displays the source address and destination address of the packet [15-16], the protocol and the packet information like length. The below figure specifies the network traffic analysis. In figure 6 and figure 7analyzing of data packet is done.

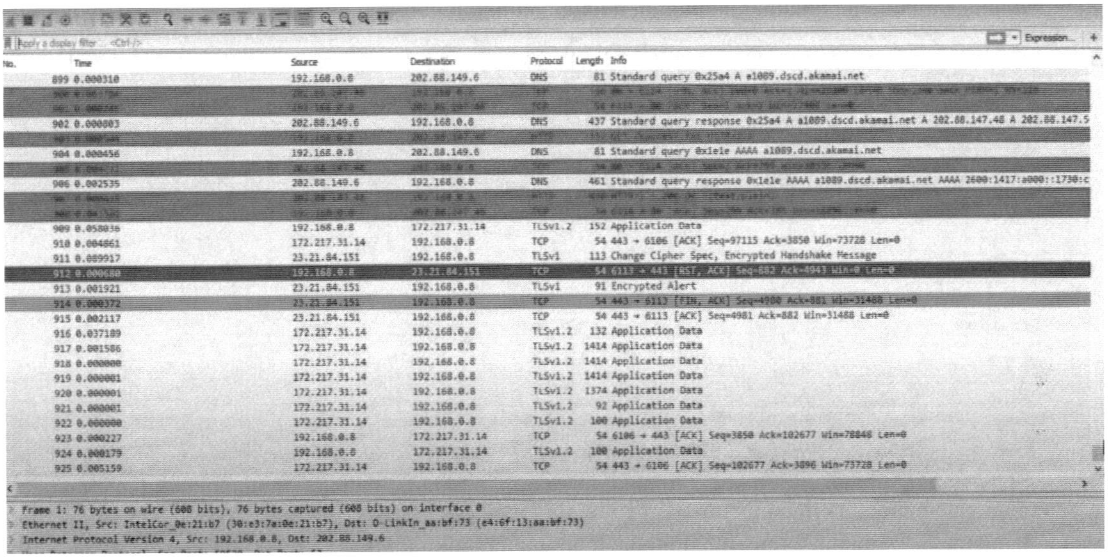

Figure. 6. Analysis of data packets

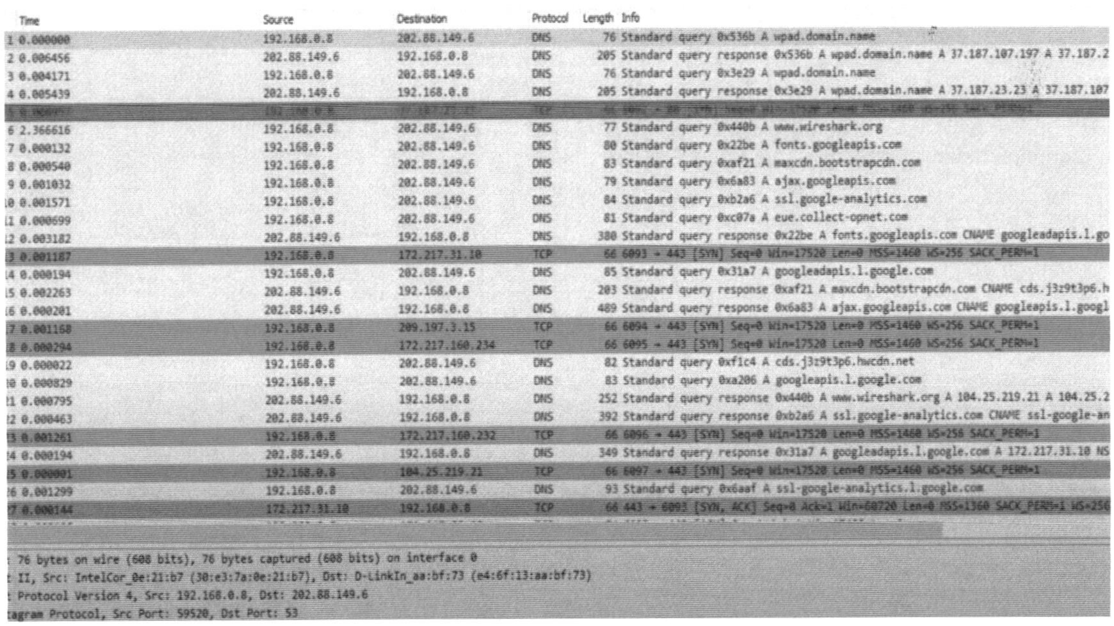

Figure. 7. Analysis of data packets

IOP Conf. Series: Materials Science and Engineering 1119 (2021) 012007 doi:10.1088/1757-899X/1119/1/012007

A line pattern of the packets is shown on the I/O graph, as seen in figure 8. The x-axis is the time interval per second, and the y-axis is the packets captured in every 1 seconds. In figure 9, TCP errors and delta time graph are shown. Delta time graph are difference between two packets.

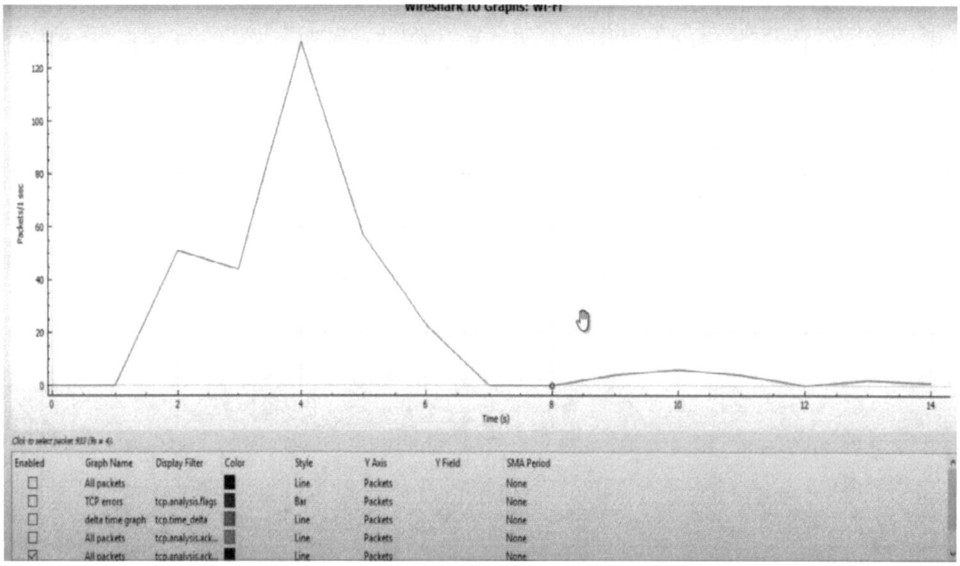

Figure 8: Wireshark I/O: ALL 950 PACKETS

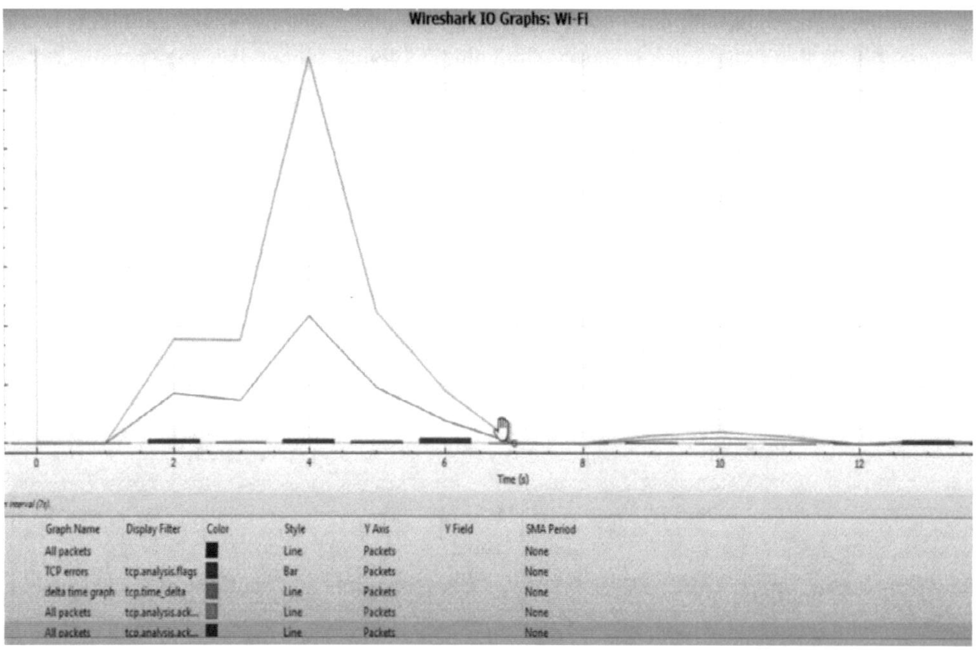

Figure 9: Wireshark I/O: TCP Errors and Delta time graph

6. Conclusion

The main issue that each agencies or companies faces in this digital scenario is network security. Network monitoring has been a vital aspect of protecting attacks. IDS plays a vivacious role in network monitoring with the help of Wireshark and Snort tools. These tools can track intrusive digital data graphically on network processes. Efficiency and security is improved by IDS and IPS network management. In this paper, Network Intrusion Detection System properties of Snort which is used for intrusion detection by making use of predefined rules and alerting the user by directing alert messages. The SNORT tool creates log file that entails data packets along with their alert messages. The recorded log file is then exported to Wireshark so as to examine the captured data packets. This Wireshark tool generates each detail about the packets of the log file. It provides minute details about frames, internet protocols, Ethernet, protocol hierarchy etc. An I/O graph outlined the flow of the packet which demonstrations the total traffic which is further measured in either bytes or packets per second.

References

[1]. Rashmi Hebbar, Mohan K., "Packet analysis with Network Intrusion Detection System" International Journal of Science and Research (IJSR), vol. 4, Issue 2, February (2015).
[2]. V. Ndatinya, Z. Xiao, "Network Forensic Analysis using Wireshark", International Journal of Sensor Networks, Vol. 10, No. 2, (2015).
[3]. M. Ahmad, K. Ullah, "Snort Implementation with Wireshark HelpingAvoiding DDOS Attack in the Cloud Computing", International Journal of Computer Science and Information Security, Vol. 15, No. 1, January (2017).
[4]. Qing-Xiu Wu, "The Network Protocol Analysis Technique in Snort", International Conference on Solid State Devices and Materials Science, (2012).
[5]. Saboor, M. Akhlaq, B. Aslam, "Experimental Evaluation of Snort against DDOS Attacks under different Hardware Configurations", 2nd National Conference of Information Assurance, (2013).
[6]. Alia yahia, Eric Atwell, "Evaluation of Capabilities of Wireshark as Intrusion Detection System", Journal of Global Research in Computer Science, Volume 9, No. 8, August 2018.
[7]. Xiaojin Hong, Changzhen Hu, Zhigang Wang, Guoqiang Wang and Ying Wan, "VisSRA: Visualizing Snort Rules and Alerts," In Proc. Of Fourth International Conference on Computational Intelligence and Communication Networks (CICN), pp.441-444, 2012.
[8]. A. EL-Dia Riad, I. Elhenawy, A. Hassan and N. Awadallah, "Using Jquery with Snort to Visualize Intrusion," IJCSI International Journal of Computer Scince Issues, pp.486-491, 2012.
[9]. Jinsheng Xu, Jinghua Zhang, Triveni Gadipalli, Xiaohong Yuan and Huiming Yu, "Learning Snort Rule by Capturing Intrusions in Live Nerwork Traffic Replay," Proceedings of the 15th Colloquium for Information Systems Security Education (CISSE), pp.145-150, 2011.
[10]. Mohammad Dabbour, Izzat Alsmadi and Emad Alsukhni, "Efficient Assessment and Evaluation for Websites Vulnerabilities Using SNORT," International Journal of Security and its Applications, pp.7- 16, 2013.
[11]. Guozi, S.U.N., Jiang, W., Yu, G.U., Danni, R.E.N., Huakang, L.I.: DDoS attacks and flash event detection based on flow characteristics in SDN. In: 2018 15th IEEE International Conference on Advanced Video and Signal Based Surveillance (AVSS), pp. 1–6 (2018)
[12]. https://snort.org/. Accessed 28 Dec 2019
[13]. Sakshi Singh, Suresh Kumar, "Capabilities of Wireshark as intrusion Detection System", International Journal of Recent Technology and Engineering, vol. 8, Issue 5, January (2020).
[14]. Sakshi Singh, Suresh Kumar, "Qualitative Assessment of Digital Forensic Tools" Vidyabharati International Interdisciplinary Research Journal, vol. 10, Issue 1, March (2020).
[15]. Snort_manual.pdf
[16]. https://www.wireshark.org/docs/wsug_html_chunked/ChAdvExpert.html

Analysing Movie Success Based on Machine Learning Algorithm

Ashutosh Shankhdhar[1], Vinay Agrawal[2], Vikram Rajpoot[3*]
[1, 2, 3]Dept. of Computer Engineering and Applications GLA University, Mathura U.P, India
[3*]vikram.rajpoot@gla.ac.in

Abstract The Film Industry is producing several hundreds of movies paving the way for the United States for third position in the list of giants of film Industry of the whole world [1]. The expenditure on these movies are of the ten to eleven figures or we can say thousands of millions of dollars, which ensure their box office success, is absolutely essential for the survival of the industry. Predicting what sort of movies are going to earn more and what type is going to come up short before the release, may profit the houses extraordinarily because it will also empower them to work on their promoting efforts which itself need numerous dollars, appropriately[2]. Furthermore, it would likewise assist them with knowing when it is generally suitable to deliver a film by taking a look at the general market. Along these lines, the examination of film achievement is critical to the business. ML (Machine Learning) algorithms [3] calculations broadly need to make expectations like development inside the stock trade, interest for items, nature of tumors, and so on.
Keywords- Data Wrangling, Data Cleaning, Exploratory Data Analysis.

1. Introduction

In the US just, the 2000s of movies are delivered per annum. Since the 1920s, the American Movies have earned additional cash per as compared to the other nation [4]. Films in America might be a multi-billion dollar industry where every single movie acquires over a billion dollars. Certain famed production houses control the maximum part of this business, where billions of dollars are spent in promotion only. Promotion and ads expenses are more intense than the whole spending plan of the movies. Most of the times this investment leads to big loss to the makers [5]. W.B.[Warner Brothers], one of the giants of creation houses, had a drop in their incomes a year ago, despite the swelling and subsequently the expanded number of flicks delivered. In the event that it had been some way or another conceivable to see heretofore the probability of accomplishment of the newly released films, the assembly houses could change the let out of their movies to increase the greatest benefit. they may utilize these findings to ensure the outcome i.e when the market is profitable and when it's not. It paves the way for a desperate requirement for such programming to be created. A lot of scholars have attempted to achieve this objective of study film incomes. The paper is coordinated as follows. Area 2 portrays the function of information set assortment and pre-processing in the information handling. Execution has been examined in segment III. Results have appeared in segment IV and Section V finishes up the paper.

2. Dataset Collection and Processing

2.1 Dataset Collection

Dataset was initially from Kaggle, and given by Udacity. As indicated by Kaggle's presentation page, the information contains data that is given from The Movie Database (TMDb)[6]. It gathers 5000+ movies' fundamental movie data and film matrices, having user rating, revenue, and popularity ratings. These measurements are frequently observed as to how fruitful these films may be. The film's fundamental data contained things like the cast, chief, watchwords, runtime, classes, and so on (Comment: The dataset on the Kaggle page might be refreshed to a newer version.) And the entire dataset term covers from 1960 to 2016.

Figure. 1. Data Definition

2.2 Data Pre-processing

The dataset we get is very mislaid, noisy, and conflicting information because of the big size and different origin, various sources [7]. We generally utilized TMD Dataset. The principal issue with datasets was mislaid fields. To remove this mislaid field issue so we embraced a way that utilizes both median and mean as a focal inclination. Then eliminated duplicated things.

3. Data Integration and Transformation

The data that we got from three unique assets Wikipedia, Rotten Tomatoes, and IMDb[8] were coordinated into one TMDb data set. So that the coordinated information is changed or united so the analysis could be effective and simpler. Dataset is blended in with both nominal and numeric characteristics. We utilized the proportion of central tendency of Box office income to change comparing nominal characteristics over to numerical.

Table.1. Features for characterising the movie attributes

Types	Characteristic
Nominal values	original_title, cast, director, overview, genres, release_date, production_companies
Numerical values	id, imdb_id, popularity, budget, revenue, runtime,vote_count, vote_average,release_year, budget_adj, revenue_adj

4. Implementation

4.1 Data Wrangling

Data wrangling is a process of taking and standardizing imperfect and disorganised raw data so that you can easily access and analyze it. Data wrangling, now and then referenced as data munging, is a way which is used for modifying and planning given dataset from one "useless" dataset structure into another form with the aim of making it more appropriate and necessary data form for an assortment of downstream purposes, for example, analytics[6].It also includes mapping data fields from source to destination. A data wrangler might be the one that carries out these transformation operations

4.2 Data Cleaning

Data cleaning is the method of updating data to make sure that it is free of errors and incorrect details. Often known as data cleaning, this includes finding incomplete, wrong, irrelevant, missing and dirty sections of the dataset and then removing or cleaning up the dirty parts of the data. Data cleaning is about finding the best approach to amplify an information set's exactness without essentially erasing information [7]. Data can be cleaned by eliminating or changing data that are incomplete, irrelevant, duplicate, incorrect, or inappropriately formatted.

Months with highest number of movie release in all years –

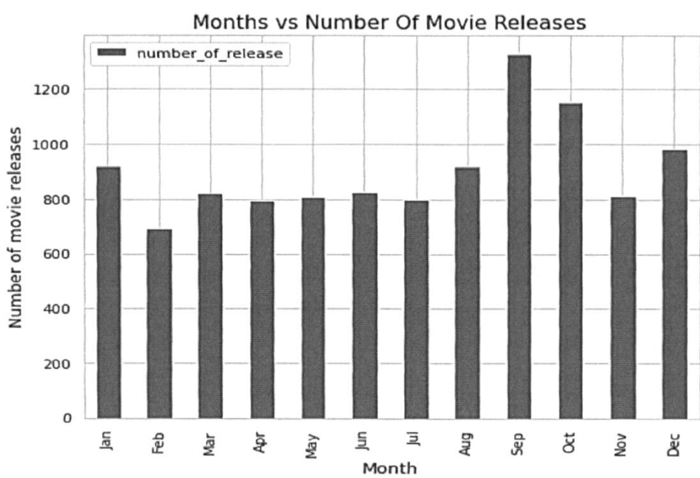

Figure.2. Month vs No. of movies releases

4.3 Data Analysis

Analysis can be considered as more of a process in order of cleaning, and transforming data to get valuable information for industry dynamics. The focus of Analysis of data is to extricate helpful information from data and settle on the choice dependent on the data analysis [8]. Presently something very similar an analyst accomplishes for business purposes named Data Analysis.

According to this graph in Figure 2, September and October are those months in which the highest number of movies are released.

Movie length liked by audience -

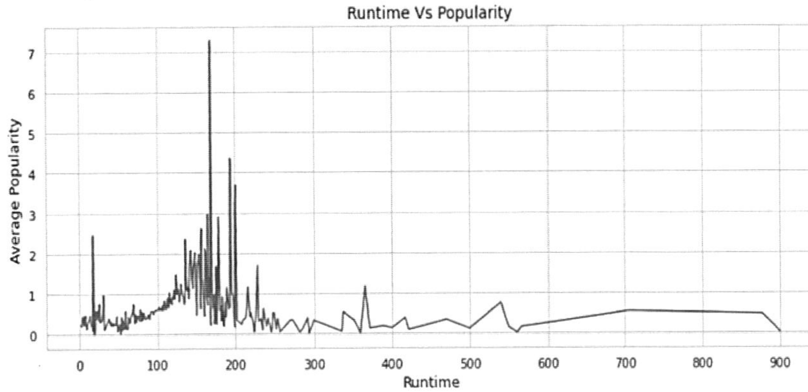

Figure.3. Runtime vs Popularity

According to the plot we can say that movies in the range of 120-200 runtime are more popular than other runtime movies. Because it is boring to see the long duration movies.

Month that made the highest average revenue

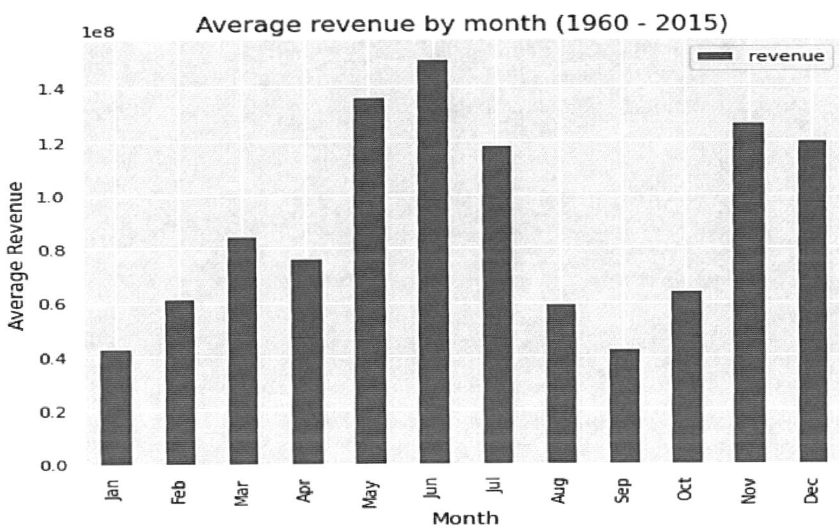

Figure.4. Revenue collected by month (1960-2015)

According to this graph May, June, November and December are those months in a year in which the highest revenue collected by the movies.

Genres with the highest release of movies –

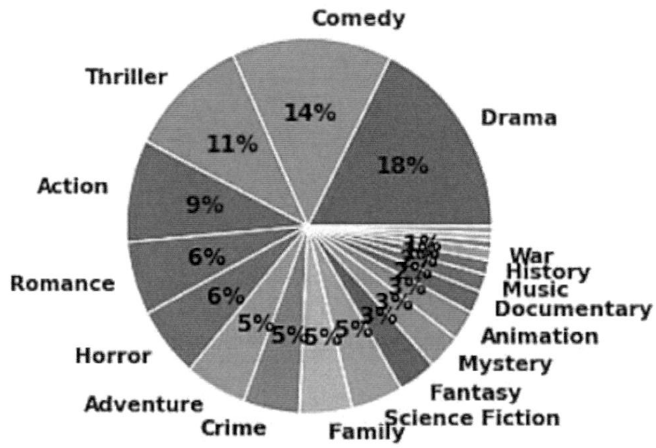

Figure.5. Popular Genres

According to this plot Drama (4761) genre has the highest release of movies followed by Comedy (3793) and Thriller (2908).

Correlation of properties associated with movies with high revenue-

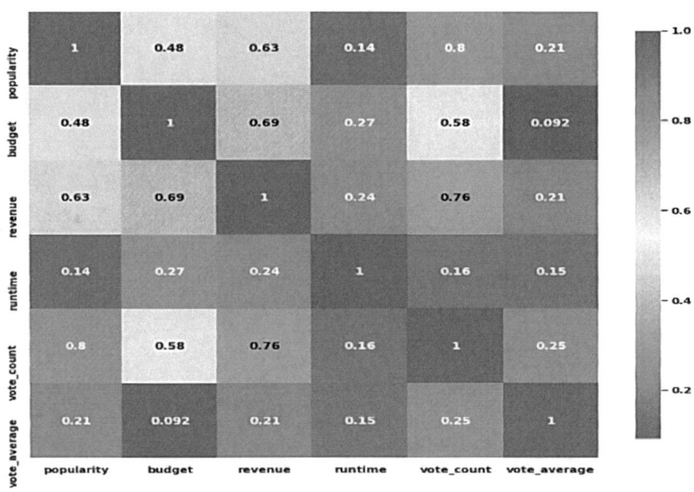

Figure.6. Properties associated with successful movies

5. Results

Budget vs Revenue: After analysing the data, it was noticed that with increase in budget, the revenue amount increases proportionally with a positive correlation (0.68) This implies there is a decent likelihood that movies with higher investment bring about more revenues.

Profit vs Budget: The research conveyed that with a profit growth budget goes in hands with profit amount with a positive correlation (0.53) This implies there is a decent possibility that movies with higher investment bring about more prominent Profit.

Release Year vs Vote Average: After investigating the findings this can be concluded that release year and voting does not affect the movie success and the same is conveyed by negative correlation (-0.11). This implies that movie ratings don't rely upon the delivery year.
Popularity vs Profit: When the outcome was analysed then it was perceived that as the newly released movie gains popularity the profit comes as a bonus providing a positive correlation (0.61). It means that movies with high popularity tend to earn high profits.
Popularity vs Revenue: In this research it was detected that popularity contributes directly towards revenue as is shown by the positive correlation (0.68). Hence it can be derived that if the popularity of movies is high then the revenue too is going to rise up.

Vote Average vs Revenue: The correlation between revenue and vote average is 0.2069. So vote average is not highly related to the revenue.

Runtime vs Revenue: The correlation between revenue and runtime is 0.2378. So runtime is not highly related to the revenue.

6. Conclusion and Future Work

After analysing the whole dataset a lot of interesting findings came into light, which contains such plentiful information that It can dig out the properties of successful movies as well as different kinds of matrices so that results can be cross-compared. A basic data analysis process was performed using the tactics learnt so far.
There were four popular genres: drama, action, comedy and thriller and Drama came out to be the most popular genre followed by action, comedy and thriller. 2014 was the year in which production of movies was at its peak. The audience prefers short movies over lengthy movies, where the average runtime of movies is decreasing each year. May, June, i.e. summer vacation and November, and December winter vacation are the months when movies earn large profit, high popularity proving to be the most suitable month for releasing to ensure maximum earnings. Thus if a production house wants to earn more profit then these findings must be taken into consideration strictly. High revenue paves the way for greater budget. Warner Bros, Universal Pictures and Paramount Pictures production companies begged the top positions and have more lifetime profit than other production companies. Movies with higher budgets have shown a corresponding increase in revenues.

In future this research is going to further advance towards an application or an app which will generate a report after analysing the previous trends i.e by taking into consideration information like what genre was liked most and what length of movie was preferred. Was it in controversy etc. In the app the producer will fill all the info like budget, releasing month, genre, and important factors and based on those factors the report will predict whether the movie is a success or not.

7. Limitations

It's not a 100 percent guaranteed solution that this formula is going to work, but it shows us that we have a high probability of making high profits if we had similar characteristics as such. If we release a movie with these characteristics, it gives people high expectations from this movie. This was just one example of an influential factor that would lead to different results, many have to be taken care of. During the data cleaning process, I split the data separated by '|' within records for simple parsing during the investigation stage. This expands the time taken in calculating the outcome

References

[1] Ahmad, J., Duraisamy, P., Yousef, A., & Buckles, B. (2017, July). Movie success prediction using data mining. In *2017 8th International Conference on Computing, Communication and Networking Technologies (ICCCNT)* (pp. 1-4). IEEE.

[2] Joshi, A. M., & Hanssens, D. M. (2009). Movie advertising and the stock market valuation of studios: a

ase of "great expectations?". *Marketing Science*, *28*(2), 239-250.
[3] Bonaccorso, G. (2017). *Machine learning algorithms*. Packt Publishing Ltd.
[4] Sklar, R. (2012). *Movie-made America: A cultural history of American movies*. Vintage.
[5] Simonoff, J. S., & Sparrow, I. R. (2000). Predicting movie grosses: Winners and losers, blockbusters and sleepers. *Chance*, *13*(3), 15-24.
[6] Kaggle datasets. (https://www.kaggle.com/tmdb/tmdb-movie-metadata)
[7] Jiawei Han, Micheline Kamber, Jian Pei, Data Mining Concepts and Techniques, 3rd ed.MA: Elsevier, 2011, pp. 83-117
[8] The International Movie Database (IMDb). https://www.imdb.com/
[9] Kandel, S., Heer, J., Plaisant, C., Kennedy, J., Van Ham, F., Riche, N. H., ... & Buono, P. (2011). Research directions in data wrangling: Visualizations and transformations for usable and credible data. *Information Visualization*, *10*(4), 271-288.
[10]Dasu, T., & Johnson, T. (2003). *Exploratory data mining and data cleaning* (Vol. 479). John Wiley & Sons.
[11] Brandt, S. (1998). *Data analysis*. Springer,.

A Study on Detection Of E-Coli Bacteria in Drinking Water using E-Nose System

Pa. Suriya[1], K. Naveenkumar[2], S P Sangeetha[3], R Divahar[4], S. Vijaykumar[5]

[1]Research scholar and Assistant Professor, Department of Civil Engineering, Aarupadai Veedu Institute of Technology deemed to be Vinayaka Missions Research Foundation, Chennai – 603104, India.
[2]Research scholar and Assistant Professor, Department of Civil Engineering, Aarupadai Veedu Institute of Technology deemed to be Vinayaka Missions Research Foundation, Chennai – 603104, India.
[3]Professor and Vice Principal, Department of Civil Engineering, Aarupadai Veedu Institute of Technology deemed to be Vinayaka missions Research Foundation, Chennai – 603104, India.
[4]Associate Professor and Head, Department of Civil Engineering, Aarupadai Veedu Institute of Technology deemed to be Vinayaka missions Research Foundation, Chennai – 603104, India.
[5]Undergraduate Student, Department of Civil Engineering, Aarupadai Veedu Institute of Technology deemed to be Vinayaka missions Research Foundation, Chennai – 603104, India.
Mail: [1]pasurya1@gmail.com

Abstract. E- nose system is mainly used to ensure the qualitative analysis of microbes accumulated in the water and air. It is an authoritative tool to evaluate the odour compounds during the quality control process. Our aim of the project is to use the E-Nose system for finding the E-coli Bacteria in the public water supply sources for drinking purposes. As the instrument for detecting E-coli Bacteria is not available, and hence we construct an instrument with a sensor having computer connectivity. This work is deliberated to measure the microbial contamination in water segregation system in metropolis and municipalities. However, the range of limits to all factors are associated to sensor with respect to the conventional methods of micro-organism detection. The collection of water from pond at sirudhavoor village in Chengalpattu district. In this sensor system, using an embedded PIC microcontroller is used to detect and quantify the microbial contaminants. The results of this analysis will give the details of CO_2 and its bacterial density in drinking water.
Keywords. E-Nose system, E-Coli Bacteria, Drinking water, Microbial contaminants, Biosensor

1. Introduction

Water is a transparent fluid which forms the world's streams, lakes, oceans and rain and is the major constituent of the fluids of living things. As a chemical compound, a water molecule contains one oxygen and two hydrogen atoms that are connected by covalent bonds. Water covers 71% of the Earth's surface. It is vital for all known forms of life. On Earth, 96.5% of the planet's water is found in seas and oceans, 1.7% in groundwater, 1.7% in glaciers, a small fraction in other large water bodies, and 0.001% in the air as vapour. Only 2.5% of the Earth's water is freshwater and 98.8% of that water is in ice (excepting ice in clouds) and groundwater. Less than 0.3% of all freshwater is in rivers, lakes, and the atmosphere, and an even smaller amount of the Earth's freshwater (0.003%) is contained within biological bodies and manufactured products. Safe drinking water is essential to humans and other life forms even though it provides no calories or organic nutrients. Access to safe drinking water has improved over the last decades in almost every part of the world, but approximately one billion people still lack access to safe water and over 2.5 billion lack access to adequate sanitation.

A report, issued in November 2009, suggests that by 2030, in some developing regions of the world, water demand will exceed supply by 50%. Water plays an important role in the world economy, as it functions as a solvent for a wide variety of chemical substances and facilitates industrial cooling and transportation. Approximately 70% of the fresh water used by humans goes.

2. Literature Review

Wedge et al. 2009 has used electronic noses to detect vapour using sensor in chemistry field. E-noses were comprising with series of organic field-effect transistors (OFETs). They are predominantly appreciated in the scope and assortment of the material which they provide concerning analyte binding. These techniques used to analyse the data using Genetic Programming (GP) and dealt with OFETs, which detect airborne in actual field. The variety of parameters which controls, on resistance, off current and mobility which gives the sensitivity, specificity and speed of sensing. (Kott, 1965) has investigated on the advantages of counting the numbers of viruses using LD 50 method over conventional methods like MPN. The procedure of the MPN method for T2 and bulk T bacteriophages performed consistent and exact over controlled conditions. The merits of such method are to determine the few numbers of bacteriophages in large sizes of fluids. In marine water, the different varieties bulk T bacteriophages with conventional methods. The particular spasm of T bacteriophage on E. coli B permits to use samples deprived of deliberating any healing to remove another bacterium. The consequences show that secondary growth of other bacteria appears. The revealing of harmful microbes in the water is the very slow process and needs the costly equipment (Bertoldi, 2012). In actual scenario, debauched recognition is fundamental for distribution of water in proper networks. the employment of wireless biosensor in the water system to detect the worsen bacteria, which is harmful to consumers. Electronic tongue is used to analyse the microbial detection in water monitoring system to ensure the quality (Lindquist 2001). The sensor system was used to describe the water quality changes using multi-electrode virtual sensor system using signal analysis. Dhoble et al. 2013 has investigated on the food safety and security using the sensor detection in order to develop the hygiene of the consumers. This scheme is established a classical model, which is used to investigate the food items on the source behaviour of the material.

3. Design and description of E-Nose System

The emerging system called E-nose is industrialised for the computerised recognition, understanding the nature of smell, gaseous matter involved. The revealing of harmful matters presents in the form of gases with the help of this device. it is applied in the various sector such as food industries, medical field, air pollutant detection. In order to ensure the quality of the drinking water, which is free from microbial content and other foreign gases. The system consists of series of sensor which is used to measure and recognize in the different pattern shows that chemical involved in the water. Gas detection sensor using Tin oxide (SnO_2) also used to specifically analyse the methanol impurities present in the alcoholic compounds. In this paper, detection of E-coli in potable water using a carbon dioxide (CO_2) sensor. The prospective and superior quality of potable water is an elemental component for the persistence of all living organisms. The source of water is taken for the study in the local village called sirudhavoor at Chengalpattu district.

3.1.PIC Microcontroller - PIC16F877A
The PIC microcontroller is used to interface the energy measurement unit and GSM module. The PIC microcontroller used here is PIC16F877A.

3.2.LCD Module
A liquid crystal display (commonly abbreviated LCD) is a thin, plane parade stratagem made up of any numeral of tint or undistinctive pels arranged in anterior of a light source. It is often utilized in battery-powered electronic devices because it uses very small amounts of electric power.
The LCD used is HD 44780. It has been used in parallel mode, which is connected to port-D for data and port-E for control signal of liquid crystal display. The HD 44780U dot-matrix LCD regulator and chauffeur LSI parades alphamerical. It can be constructed to ambition a dot-matrix fluid quartz parade beneath the regulator of a 4-bit or 8-bit microchip. Since all the purposes such as parade RAM, character originator, and fluid quartz motorist, obligatory for lashing a dot-matrix fluid quartz demonstration are within providing on one blemish, a trifling arrangement can be interfaced with this regulator/motorist.

3.3. Hydro CO_2 Sensor
Dissolved CO_2 molecules diffuse through a specialized thin film composite membrane into the internal gas circuit leading to a detector chamber, where the partial pressure of CO_2 is determined by means of IR absorption spectrometry. Concentration dependent IR light intensities are converted into the output signal from calibration coefficients stored in firmware and data from additional sensors within the gas circuit.

3.4. Power Supply
A power supply (sometimes known as a power supply unit or PSU) is a manoeuvre or arrangement that stores electric or supplementary categories of energy to an output load or group of loads. The term is most commonly applied to electric energy provisions, fewer frequently to motorised ones, and infrequently to others. The transformer steps up or steps down the input line voltage and isolates the power supply from the power line. The rectifier section converts the alternating current input signal to a pulsating direct current. However, as you proceed in this chapter you will learn that pulsating dc is not desirable. For this reason, a filter section is used to convert pulsating dc to a purer, more desirable form of DC voltage.

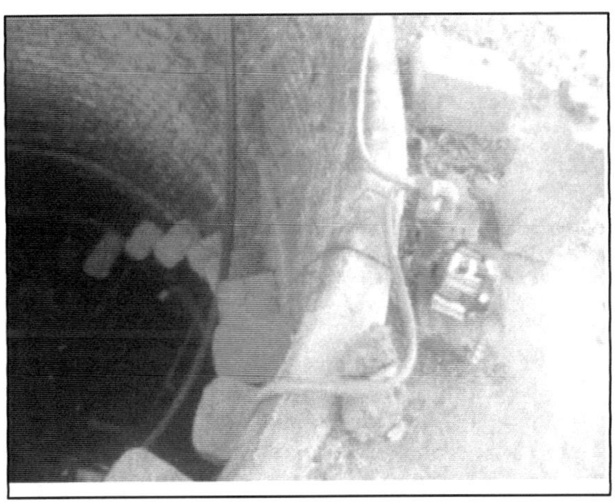

Figure 1. Well Water Sample has been collected

4. Testing of The Samples Using E-Nose System

The exposure of germs in drinking water that are destructive to healthiness, elasticities a antique synopsis of the peril of ailment from pathogens. E-coli is the furthermost copious coliform organism contemporary in the human and animal intestine, and can be existing in faces causing a main concern to health.

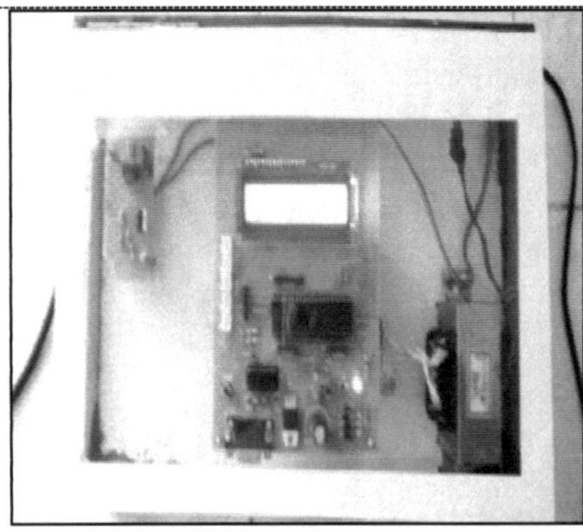

Figure 2. Instrument has been set to ON position and calibrated

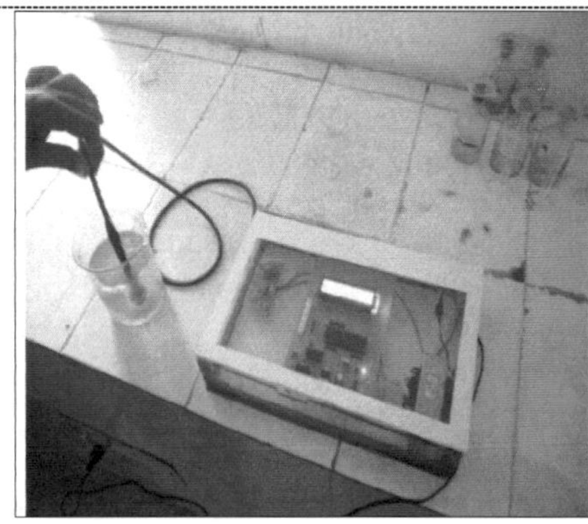

Figure 3. Sample is tested and the readings are displayed in the display screen

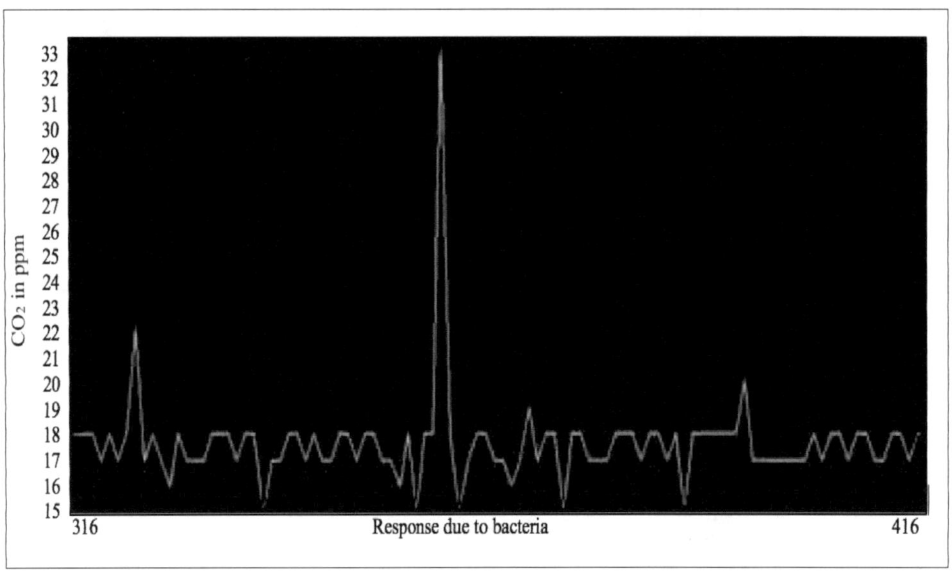

Figure 4. Pure Drinkable water having 18ppm of CO_2

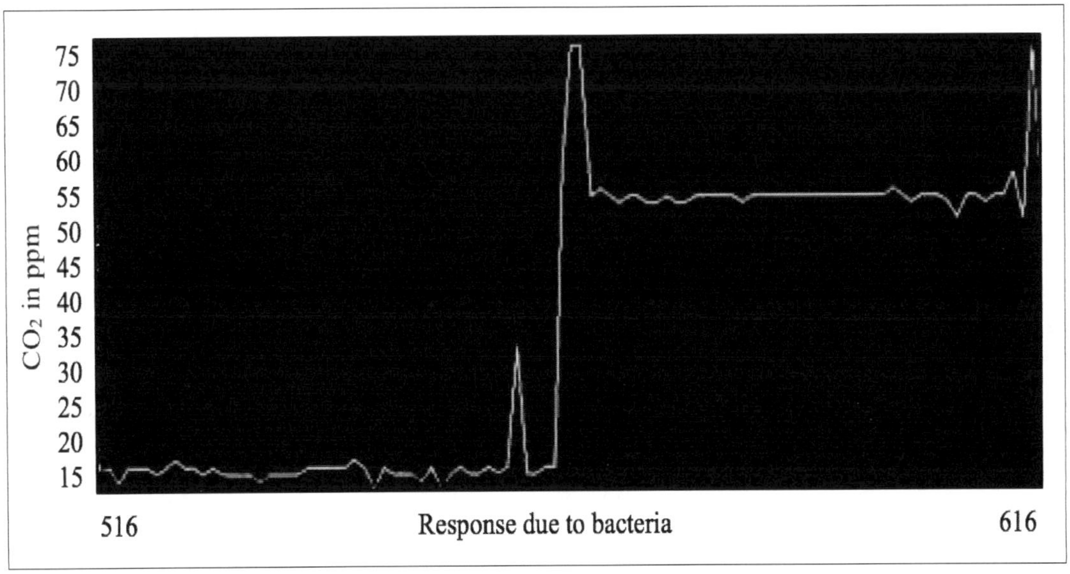

Figure 5. Pure Drinkable water having 52 ppm of CO_2

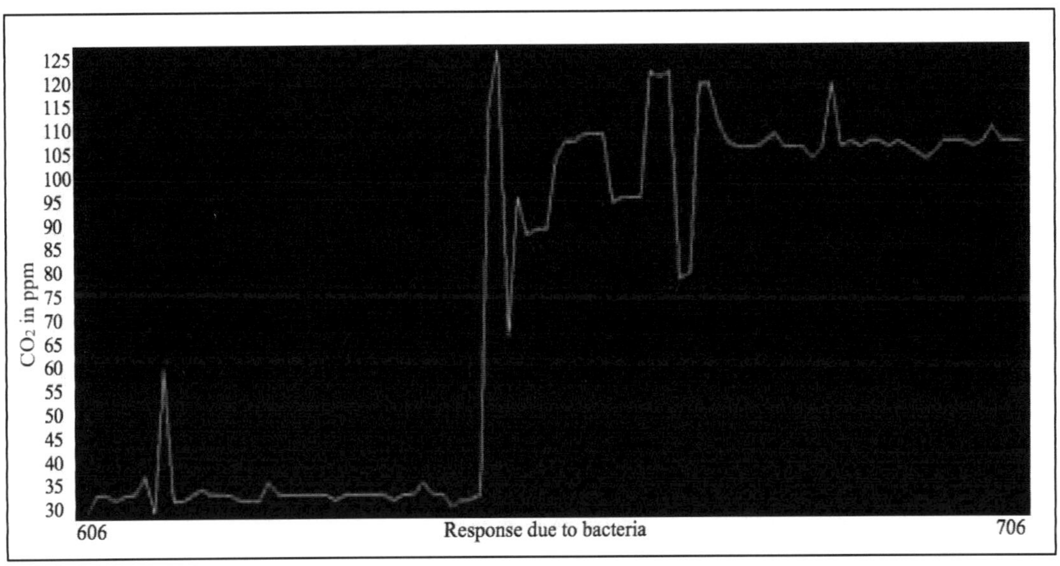

Figure 6. Non Drinkable water having 110 ppm of CO_2

The results have shown that microbes present in the drinking water analysed accurately with the help of sensors. The Threshold limit of contaminants were measured and qualitatively distributed.

5. Results and Discussion

As per Indian Standards IS: 10500 – 1991 drinking water has its permissible limits to ensure safety to use the water for drinking purpose. The parameters which is to be tested is turbidity, taste, PH value, presence of hardness and presence of E-Coli bacteria in drinking water. For the purpose of detecting the E-Coli bacteria an instrument with a sensor having computer connectivity is constructed. The system of testing the drinking water is termed as E-Nose system. The prearrangement parades the eminence of drinking water and demonstration the percentage level of microbial contaminants. This work is proposed to distinguish organic water toxins like coliform group of bacteria in drinking water. Yet, a numeral of restrictions is allied over traditional methods of microbial detection.

6. Conclusion

E-nose device is used to found the effectively quantifying the microbial contaminants in the drinking water. the survival of bacterial in water is based on the presence of carbon dioxide content. This study is aimed at qualifying the carbon dioxide content. If the limits of carbon dioxide content in the range of 15 to 50 ppm, then it is considered as a potable water. When the limit exceeds the 110 ppm, it is not fit for drinking purposes. The results were shown that bacterial count and also comparing with the traditional method with sensor response. This system results shows the quality of water and the range of microbial contaminants expressed in percentage. Future studies would be to improve the system by assimilating radical chemical similarity sensors for identifying chemical impurities and hardness of water.

7. References

[1] Bertoldi, A., 2012. Modelling studies for the detection of bacteria in Biosensor Water Distribution Networks.

[2] Wedge, D.C., Das, A., Dost, R., Kettle, J., Madec, M.B., Morrison, J.J., Grell, M., Kell, D.B., Richardson, T.H., Yeates, S. and Turner, M.L., 2009. Real-time vapour sensing using an OFET-based electronic nose and genetic programming. Sensors and Actuators B: Chemical, 143(1), pp.365-372.

[3] Dhoble S.B. and Choudhari N.K., 2013. Neuron Model to Analyze the Infection behavior in sample food material, International Journal of Engineering Research and Technology (IJERT), 2(11), pp.421-424.

[4] Kott, Y., 1966. Estimation of low numbers of Escherichia coli bacteriophage by use of the most probable number method. Applied microbiology, 14(2), pp.141-144

[5] Lindquist, M. and Wide, P., 2001, May. Virtual water quality tests with an electronic tongue. In IMTC 2001. Proceedings of the 18th IEEE Instrumentation and Measurement Technology Conference. Rediscovering Measurement in the Age of Informatics, 2, pp. 1320-1324.

Performance Improvement of Circular Micro-Strip Patch Antenna for WCS

S. Kannadhasan[1], **R. Nagarajan**[2]
[1]Assistant Professor, Department of Electronics and Communication Engineering, Cheran College of Engineering, Tamilnadu, India
[2]Professor, Department of Electrical and Electronics Engineering, Gnanamani College of Technology, Tamilnadu, India
[1]kannadhasan.ece@gmail.com

Abstract: - The proposed a circular micro-strip antenna (CMSA) with a circular slot in this paper, that solves the current narrow bandwidth dilemma. The parametric analysis reveals that only with the right choice of parameters would wideband features be achieved with a smaller antenna. The proposed antenna is built at 27.5GHz with wider bandwidth and tremendous return failure. This antenna can also be used extensively for embedded applications with a large data rate. This antenna can be physically implemented with embedded devices that enable it to function in a broader variety of applications.
Keywords: - WCS, CMPSA, Reflection Coefficient, Gain and Radiation Pattern

1. Introduction

GPS (global positioning system) receivers are now commonly used predominantly for Wireless Communication System (WCS) such as automobiles, aircraft, ships, and cellular communication through satellite navigation. Patch and quad helix are the two most common antenna styles used in GPS receivers. Thanks to their light weight, ease of mounting, single hemispheric radiation pattern and low profile, micro-strip patch antennas are ideal for a GPS receiver antenna. This is focuses on the design and produce of a new form of circular micro-strip patch antenna (CMSPA) inset feed. Enhanced patch antenna characteristics can be accomplished by adding a circular slot at the middle of the ground plane. Focus has been put on optimizing the 1.8GHz antenna configuration by using CST-Microwave Studio simulation tools. It is observed from simulation and calculated outcomes that the proposed patch antenna displays better efficiency relative to a typical CMSPA [1]-[5].

One of the easiest ways to produce is the micro-strip line feed, since it is a conducting strip connected to the patch which may thus be viewed as a patch expansion. By manipulating the inset location, it is easy to model and quick to fit. Because the current at the end of a half-wave patch is low and rises in amplitude towards the middle, if the patch was fed closer to the centre, the input impedance (Z=V/I) could be minimized. Connection today plays an important position in global culture and the communication infrastructure is transitioning quickly from wired to wireless. Telecommunication is still seeking, at the lowest possible expense, to produce the highest results, reliability and productivity. In this domain, antennas constitute a fundamental element which enables electromagnetic waves to be transmitted in free space. We see many kinds of antennas that vary by cuts, geometric form, and transmitting power. A micro-strip patch antenna is a type of antenna that provides a low profile, i.e. small and simple to produce, which gives conventional antennas a great advantage. In wireless connections and other microwave uses, patch antennas are flat antennas used [6]-[10].

The micro-strip approach is a planar technique used for generating lines that transmit signals and combining those lines and radiated waves with antennas. Usually, a patch is larger than a strip and its form and dimension are significant antenna characteristics. When use MSPA, micro-strip antennas are especially ideal for large losses and surface waves in the substrate layer, since losses can often arise in the radiation when the signals are emitted by the antenna. Because of their advantages such as light weight, low volume, low expense, compatible with integrated circuits and simple to mount on the rigid surface. MSPAs are perhaps the most frequently used type of antennas today. In addition, they can conveniently be built to function in quad-band, multi-band, dual or circular polarization applications. In several commercial applications, they are important. But for realistic purposes, MSPAs have inherently narrow bandwidth and bandwidth expansion is typically needed, so several techniques have been used to expand the bandwidth. Using the dielectric constant of substrate FR4 glass epoxy 4.4 with 50 micron copper thickness, FR4 epoxy for MSAs at various frequencies such as 2GHz, 4GHz, 6GHz, 8GHz and 10GHz are design in the proposed CMSPA. Because of its low cost and easy availability, FR4 was selected for this analysis and can therefore be used for MSA array prototyping [11]-[13].

Wireless connectivity has been part of daily life in the last few years, together with its different ways. It was this focus on wireless technology that made it possible to find small-sized and light-weight antennas. Both these ones, all of this usage of microstrip antennas fulfils the demands in contact networks, microstrip antennas are used to due to systemic simplicity, low processing costs, modest development costs, small size and simplicity of installation. The circular microstrip antenna consists of a separated by a sandwich of two parallel conducting layers with single thin substratum dielectric. It is a plain ground plane and the upper conductor Patch circular. Due to the patch antenna, it is really common day after day. its simplicity of research and production, low expense, and light weight. Microstrip patch antennas have been commonly used lately satellite communications, aerospace, radar and biomedical communications [14]-[16].

2. Modelling of Circular Patch Antenna

There are various approaches to build and evaluate a MSPA, such as hollow prototypes for transmission line patterns. Since its start, wireless and smart phone interaction has come a long way. Miniature sized antennas such as the MSA are therefore necessary for efficient communication. In this article, an appropriate circular MSA with a broad band is suggested. In this build of a circular patch is placed on the FR4 substrate material with ε_r=4.4. To have broadband operation, a circular slot is engraved on the circular patch. The antenna was designed for 27.5GHz operation and the findings showed a broader 800MHz bandwidth and a low -24.8dB return loss and an antenna gain of 8dBi. The radiating patch is positioned around the axis of symmetry with an inset feed line spanning the middle of the patch. Through removing one spherical slot, the winding of the patch is modified and a pair of rectangular slots is merged to minimize the impedance mismatching losses for wireless contact device implementations, one of which is imprinted onto the patch's non-radiating core and two are combined.

$$Wg = 6h + W \tag{1}$$

$$Lg = 6h + L \tag{2}$$

The symmetrical design of the pair of slots is attributed to the fact that it eliminates cross polarization. The slits promote the present streak and decrease the activity of the frequency. The values of each of the variables used in the design model are given in the following Table 1. The antenna design consist substrate of a single layer of thickness 1.6mm. The dielectric constant of the substrate is 4.4 and antenna is designed on FR4 as shown in Figure 1.

Table I: Antenna Design Parameters

Sl. No.	Parameter	Values
1	Operating Frequency	27.5GHz
2	Radius of Circular Patch	8.96mm
3	Substrate Dielectric Material	FR4 epoxy
4	Substrate Dielectric Constant	$\varepsilon_r = 4.4$
5	Substrate Thickness	1mm
6	Feeding Technique	Line feed
7	Ground Plane	30mm*30mm

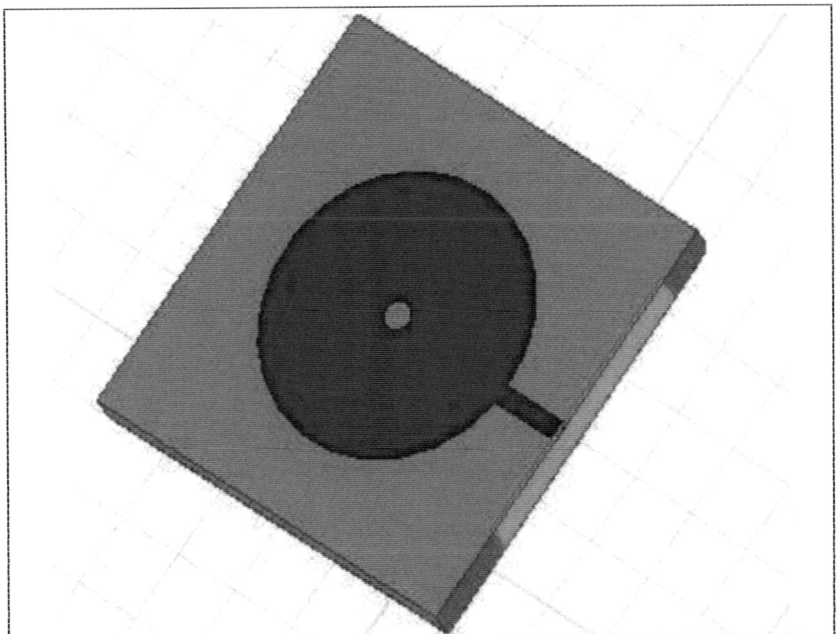

Figure 1. Proposed Structure of the Circular Patch Antenna

3. Results and Discussion

The FR4 dielectric substrate material with a relative permittivity (ε_r) of 4.4, loss tangent (δ) of 0.02 and a thickness or height of 1.6mm is included in the proposed production model. As a mix of traditional forms such as round and rectangular, the concept layout consists of a novel rounded patch. At the middle of the patch, the spherical structure is positioned with rectangular shaped strips around it. Among such rectangular patch strips, one is used as a line feed for cooking. A circular ring-shaped defect found in the partial ground comprises the ground plane. The scale of the antenna is around 30mmx30mm. The reflection coefficient parameter of the simulated effects of the proposed antenna is given in Figure 2. The 800MHz bandwidth is calculated and the estimated reflection coefficient results are recorded at -24.8dB operating frequency at 27.5GHz. There is strong consistency between the simulated and calculated values of the coefficient of reflection.

IOP Conf. Series: Materials Science and Engineering 1119 (2021) 012010 doi:10.1088/1757-899X/1119/1/012010

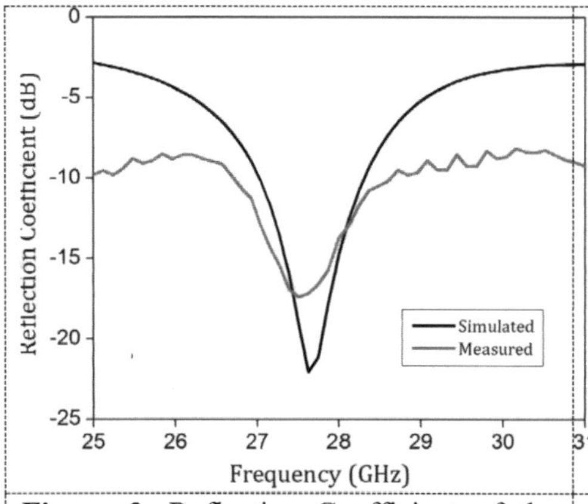

Figure 2. Reflection Coefficient of the CMSPA

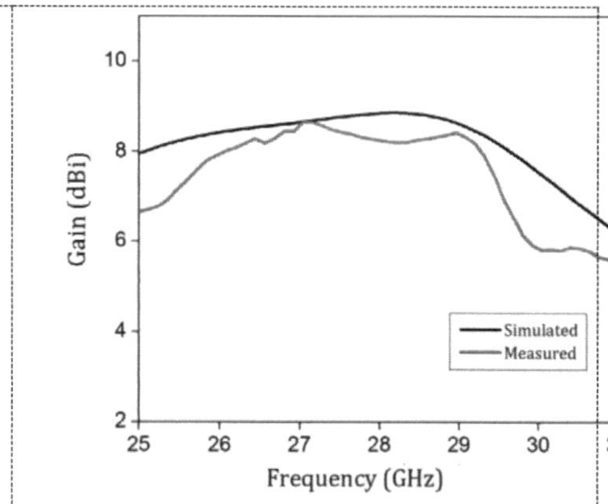

Figure 3. Gain of the CMSPA

In the CMSPA with a circular slot, the return loss is in the range of -24.8dB with a 3dB bandwidth of around 800MHz. The bandwidth is obtained with the above-mentioned patch and substratum dimension. The bandwidth obtained demonstrates that for embedded applications, the built CMSPA has high data rate applicability. The Gain Plot of the built antenna is shown in Figure 3. The benefit along the elevation plane of the proposed model is found to be almost 8dBi at 27.5GHz. HFSS is often used to simulate the gain of the proposed CPA and it is found to be about 8 dBi for the far-field pattern. Automatic GPS monitoring systems have allowed antennas to work in multi-frequency bands in recent years, including GSM, 3G, and modern LTE bands. Thanks to their light weight, ease of installation and less air-drag, several researchers emphasizes it because MSPAs are ideal for GPS use.

A circular slot in the ground plane would be added to boost the characteristics of the circular patch antenna. The new distribution of the circular patch antenna's E-Field and H-Field is seen in Figure 4 and Figure 5. At 27.5GHz, this suggested CMSPA only has one resonant frequency. Bandwidth, gain, beam width and the magnitude of reflection coefficient have increased with the implementation of the slot in the ground plane. There is a discussion of the calculated and assessed parameters of the proposed model. The features of the proposed model are simulated using high frequency structural simulator, such as reflection coefficient, gain, bandwidth, radiation trends, performance and current distribution.

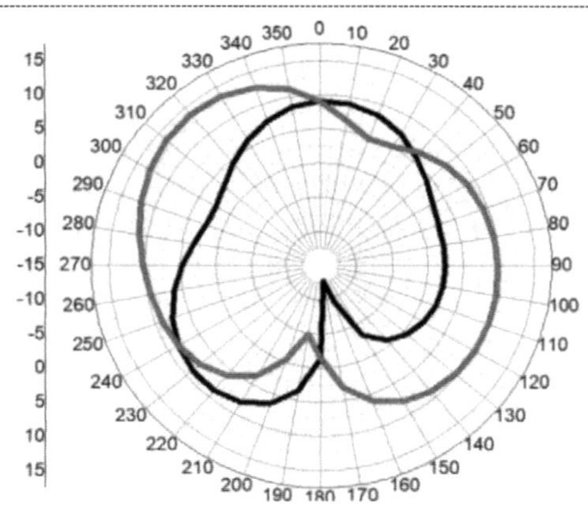

Figure 4. Radiation Pattern of CMSPA for E-Plane

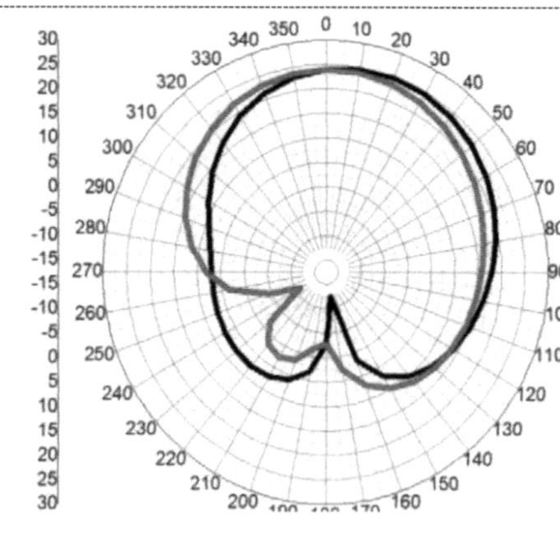

Figure 5. Radiation Pattern of CMSPA for H-Plane

4. Conclusion

Modelling and simulation of a Circular Micro-Strip Patch Antenna in the field of WSN is present in this article. The expressions of fields inside the cavity, far fields and resonant frequencies are thus obtained using the cavity model for the study of circular form of MSA. Using dielectric substrates with dielectric constant $\varepsilon_r = 4.4$ and thickness h = 1.6mm, a probe fed circular MSA is built at 27.5GHz. The radius of circular patch a = 8.96mm at these parameters. The simulated effects of CMSPA indicate strong coefficient of reflection and value of gain. For the circular patch antenna, the reflection coefficient and gain are -24.8dB and 8dBi, respectively. However, to enhance the applicability of this patch antenna may also boost the antenna characteristics utilizing various technologies.

References

1. L.K.Balaji, et al., 2018, "A survey on fractal antenna design", International Journal of Pure and Applied Mathematics, volume 120, No. 6.
2. John P. Gianviffwb. et al., 2002, "Fractal Antennas: A novel antenna miniaturization technique and applications", IEEE Antenna's and Propagation Magazine, Vol. 44, No. 1.
3. M. Kazerooni, et al., 2009, "Design and fabrication of wide band printed multi-ring fractal antenna for commercial applications", PIERS Proceedings, Beijing, China, March 23-27.
4. Zehua Zhou, et al., 2016, "A multi-band fractal antenna for RF energy harvesting", IEEE International Symposium on Antennas and Propagation, 10.1109/APS.2016.7696017
5. Raj Kumar, et al., 2012, "On the design of UWB circular fractal antenna with notched-band characteristics using W-shape slot", Journal of Microwaves, Optoelectronics and Electromagnetic Applications, Vol. 11, No.2.
6. Raj Kumar, et al., 2012, "On the design of tree-type ultra wideband fractal antenna for DS-CDMA system", Journal of Microwaves, Optoelectronics and Electromagnetic Applications, Vol. 11, No.1.
7. Tiang, J. J., M. T. Islam, N. Misran, and J. S. Mandeep, 2011, "Circular microstrip slot antenna for dual frequency RFID application," Progress In Electromagnetics Research, Vol. 120, pp-499–512.
8. Yang, Y. H., J.-L. Guo, B.-H. Sun, and Y.-H. Huang, 2016, "Dual-band slot helix antenna for global positioning satellite applications," IEEE Transactions on Antennas and Propagation, Vol. 64, No. 12, pp-5146–5152.
9. M. T. Islam, M. N. Shakib, N. Misran, 2009, "Broadband e-h shaped microstrip patch antenna for wireless systems", Progress In Electromagnetics Research, PIER 98, pp-163-173.
10. Hung, C. and T. Chiu, 2016, "Design of dual-band cavity-backed slot antenna loaded with spurline," IET Microwaves, Antennas and Propagation, Vol. 10, No. 9, pp-939–946.
11. Kandasamy, K., B. Majumder, J. Mukherjee, and K. P. Ray, 2016, "Dual-band circularly polarized split ring resonators loaded square slot antenna," IEEE Transactions on Antennas and Propagation, Vol. 64, No. 8, pp-3640–3645.
12. Rui, X., J. Li, and K. Wei, 2016, "Dual-band dual-sense circularly polarised square slot antenna with simple structure," Electronics Letters, Vol. 52, No. 8, pp-578–580.
13. S. Kannadhasan and R.Nagarajan, 2020, "Performance Design and Development of Environmentally Safe W-Shaped Structure Antenna for Wireless Applications," Journal of Green Engineering (JGE), Volume-10, Issue-9, pp-4558-4565.
14. A. Wahab, N., Bin Maslan, Z., Muhamad, W.N.W., Hamzah, N., 2010, "Microstrip rectangular 4x1 patch array antenna at 2.5ghz for wimax application", Second International Conference on Computational Intelligence, Communication Systems and Networks (CICSyN).

15. Dundar, D. Uzer, S. S. Gultekin, M. Bayrak, 2012, "Effects of microstrip feed line width on 1×4 rectangular microstrip antenna array electrical parameters and estimation with artificial neural networks", Progress In Electromagnetics Research Symposium Proceedings.
16. B.-K. Angand, B.-K. Chung, 2007, "A wideband e-shaped microstrip patch antenna for 5–6 ghz wireless communications", Progress In Electromagnetics Research, PIER 75, pp-397–407.

Study of Cloud Computing Business Framework by Pre-cloud recognition

N Tiwari[1], **N Sharma**[2]
[1,2]IIS (Deemed-to-be University), Jaipur, India.
[1]Neha.08sep@gmail.com

Abstract: The current study strongly analyzes the factors that determine the acceptance of cloud companies (SASS model), and this strategy to do their job. The research model is designed to explore factors that affect the use of computers, including host technology (TAM) models and other external such as organizational size and technical complexity. Data compiled from 200 companies are used to test the ideas. The results of this study show what important factors need to be considered and how they relate to each other. They highlight the needs of companies, especially those related to cloud operations, for those who want to use a real digital governance model. These technologies and technology-related applications that use virtual technology and operate in a distributed network that is accessible to sites and social networking sites. Developed and upgraded to cloud computing systems, cloud grid and computers.

Keywords: Cloud Computing, TAM, SASS, Cloud.

1. Introduction

Here introduce the paper, and put a nomenclature if necessary, in a box with the same font size as the rest of the paper. The paragraphs continue from here and are only separated by headings, subheadings, images and formulae. The section headings are arranged by numbers, bold and 10 pt. Here follows further instructions for authors. There are three main types of cloud: Duration of service depending on data acquisition and storage; Cloud Speaker is a platform (PASS) as a tool to help engineers improve applications and resources. Common benefits of integration include: reducing software and computer services. Complex use of cloud-based solutions [2], market-based formulas [9]; The power to change business processes [1]; Its ability to provide better data management and a stronger connection between information systems and management needs [2] and allows resources to divert key business functions due to reduced costs and IT-related operational costs [3]

In addition, there are challenges and problems in controlling access to cloud computing, such as long-term security and protection against unauthorized access [10], ignorance of service providers' privacy, and misunderstandings between system providers and clouds, technical problems and financial difficulties in some companies may require appropriate technical solutions and investment in information technology for employees, especially in small and medium enterprises [11] and infrastructure security such as performance and quality. Process Assessment Process [9], as well as surrounding problems at levels related to the problem of their integration and customization [10].

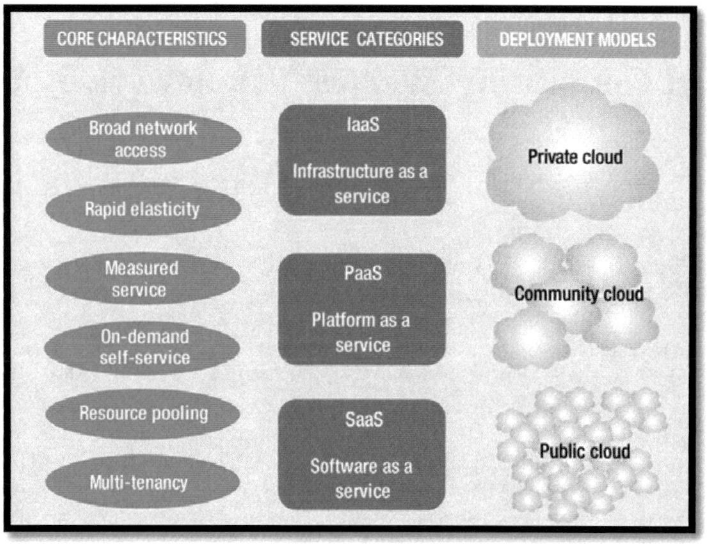

Figure1: Service Categories of cloud

2. Literature Review

New species. Various technological advances can be found in the literature that provides an effective framework for determining critical or variable factors that contribute to the user's acceptance of ICD and its conduct in organization [12]. The same is true of Fishbean and Ajgen [13] Theory of Recent Action (DRA). This view, adopted in many fields, explains that the method by follows a given action is related to the purpose of adhering to the stated behavior. In addition, other people's opinions about certain technologies often influence a person's intentions to use the same. Ajgen later introduced The Planned Behaeve Theory, adding to TRA's confidence in the management and practice and control of behavior.

In addition to the adoption of the cloud system and its use, there have been many recent studies in the field of information that use TAM to demonstrate the acceptance of various technology. Many of them can incorporate other external variables into the actual TAM which is considers to attract research. Therefore, courses are most commonly used in the ERP system with companies [14], e-commerce applications, digital file systems, online banking, public gambling, and research forums [16], in many fields. At present, the structure of the TOE (Technology-System-Innovation [17] indicates that the adoption of new technologies is influenced by three factors: the organizational environment.

TOE is one of the most widely distributed structures in the ICT adoption industry [18]. It provides a comprehensive view of the acceptance of technologies and their use and predicts the impact of price options and the subsequent increase in factors affecting business, the basic structure of the TOE is unclear and some of the errors found in these three conditions are different in different studies.

In addition to the adoption of the cloud system and its use, there have been many recent studies in the field of information that use TAM to demonstrate the acceptance of various technology. Many of them can incorporate other external variables into the actual TAM which is considers to attract research. Therefore, courses are most commonly used in the ERP system with companies [14], e-commerce applications, digital file systems, online banking, public gambling, and research forums [16], in many fields. At present, the structure of the TOE (Technology-System-Innovation [17] indicates that the adoption of new technologies is influenced by three factors: the organizational environment.

TOE is one of the most widely distributed structures in the ICT adoption industry [18]. It provides a comprehensive view of the acceptance of technologies and their use and predicts the impact of price options and the subsequent increase in factors affecting business, the basic structure of the TOE is unclear and some of the errors found in these three conditions are different in different studies.

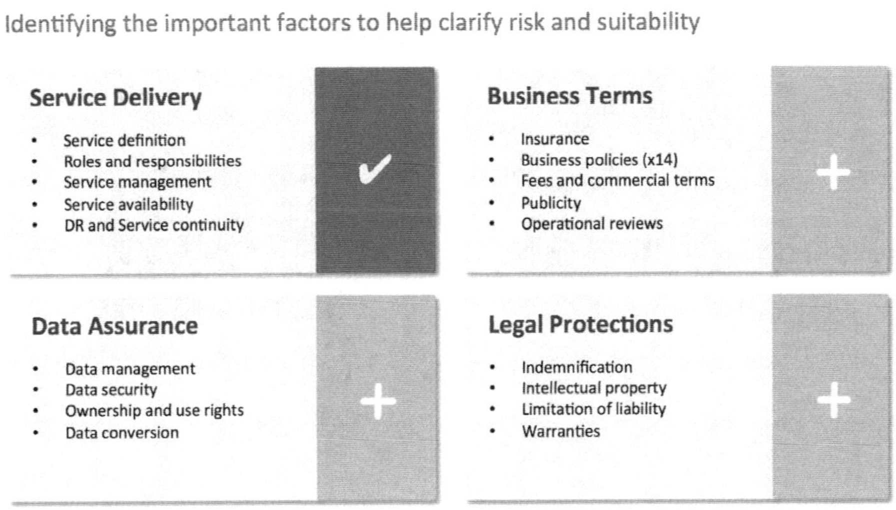

Figure 2: Danger and suitability of the cloud

3. Pre-cloud recognition.

Other ideas based on ideas and models, which focus on a specific computer area, will consider the acceptance and application of the end user's opinion. This is the position of Bhattacharjee and Park [19], who study user authorization from computer-server computing models, and Keysman and Stanowska [20], who analyzed consumer preferences in a recent survey. Finally, a recent study by Kangwar Developed an integrated model, which incorporated various experimental features and was tested using samples from 200 companies in various field in India. They suggest the power of digital content

4. ID Infrastructure Library (ITIL) Version 3 Resource Framework

The framework of ITIL V3 sets out the best practices for the management of IT services. It provides an IT management framework and focuses on improving the quality of continuous services provided [21].

- Strategic Action Plan - Provides guidence on how to design, improve and implements services management.

- Design - This is about developing and improving information technology services.
- Transformational Transformation - This process focuses on the provision of information technology service.
- Operation Service Function - This ensures that IT services are delivered efficiently and effectively.
- Sustainable Service Development - This process focus on improved the quality of existing service.

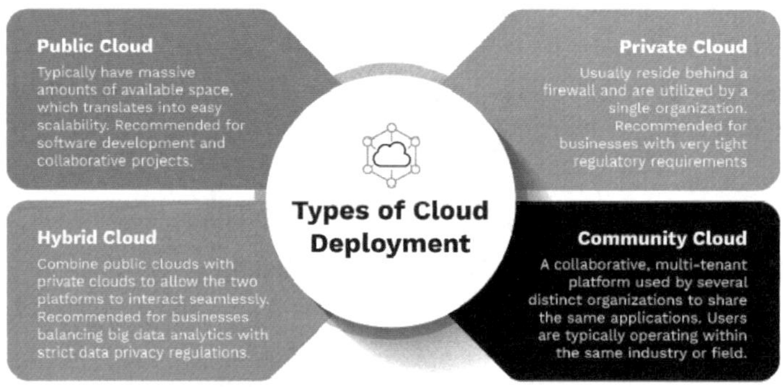

Figure 3: Various types of Cloud Deployments

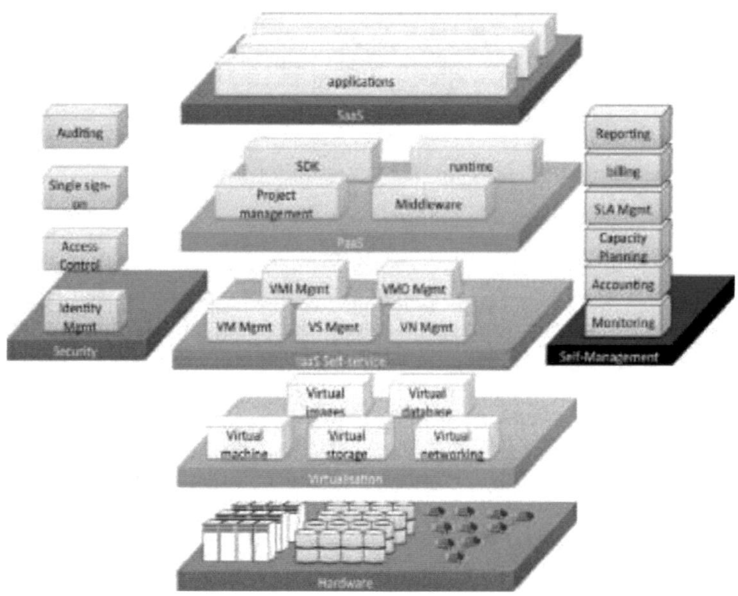

Figure 4: Application of SAAS in Virtual Storage

IOP Conf. Series: Materials Science and Engineering 1119 (2021) 012011 doi:10.1088/1757-899X/1119/1/012011

5. High relationships B/W business models & services

Projects is a widely used sector framework, which includes project management, management and planning, especially in IT-based projects. (Government Business Office, 2009) describes the project co-ordination process [22], project design and management, and what needs to be done to implement the project. Divided into manageable categories, PRINCE2 enables better resource management. This applies to Cloud Computing, because resource management is not only about Service Quality& Service Level Agreement, but also need to be considered strategically considered [23].

In addition, many projects have helped boost the economy for decades. Most of those startups were financial providers, which helped them grow and grow their businesses [24]. There were various SMEs such as HP and Cisco, which closed other businesses and expanded to include global companies with appropriate investment, integration and acquisition and integration and integration and integration of their products and services.. According to Lasonic, there are four types of businesses identified [25]:

(i) public funds;

(ii) Venture Capital;

(iii) Funding agreements and lease agreements with businesses there are researchers who support Lasonic points. Maintain their businesses.

Third, EDRO (2009) evaluates EU SME-computing and SMEs based on models of sponsorship agreements and service contracts. Finally, Weininhard et al. (2009a; 2009b) proposed a well-defined business model and aligned with the "All in One Enterprises" model.open source business models and multiple case studies [26]. They have a model of similar support agreements, such as "support agreements and services" in the Lasonic definition. They also have a model that acts as a “stand alone and is used” by vendors, users, partners, vendors and participants to share and reap similar benefits on the same platform [27]. This allows the public sector to integrate and provide a platform for sharing resources [28].

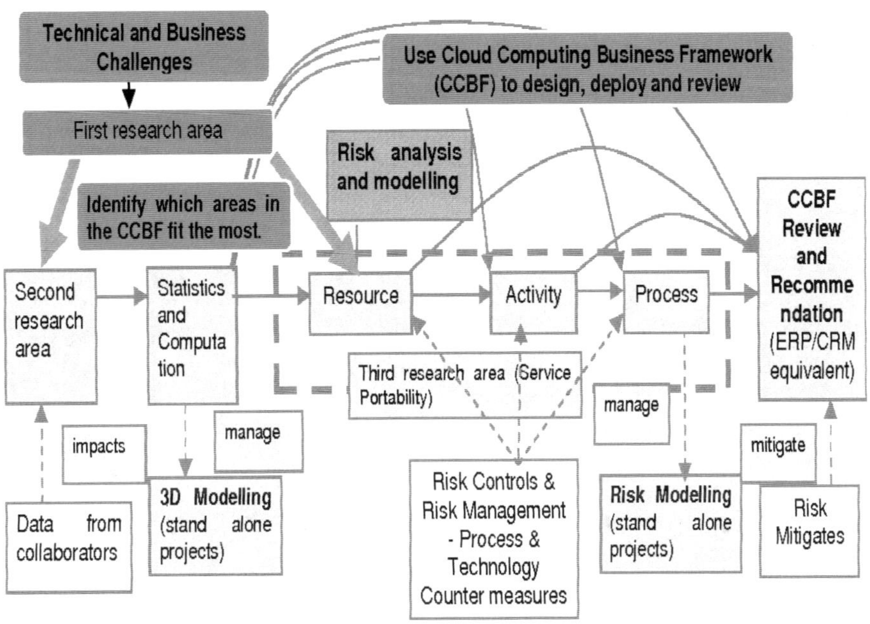

Figure 5: Overview of computing area linking both computing and business Area

5.1 Clouds for business content clouds

Apart from SLA-based research, few structures or models can show the relationship between multiple manufacturers and operators [29]. Second, there is no direct way to analyze the performance of cloud trading outside the stock market. Stock market disruption is to problems of accuracy and. There are researchers who focus on differences in the business model and the reasons why the cloud business is successful (Chow, 2009; Weinhard et al., 2009a; 2009b). But this part of the business model requires additional financial charges and additional data performance to ensure durability [30].

6. Relationships within services

Trung and Duster (2010) confirmed this, showing that they can perform continuous and completed tasks such as classification, modeling and simultaneous testing. This means that the function of each research question is considered an integral part of the structure [31] Consistent with Weinhardt. Integration of business content Chang et al. (2011a) and research issues related to planning, organizational sustainability, performance management and communication. Each site is compatible with all IaaS, PaaS and all. Each key category is defined as follows [32].

7. Communication. Communication between various services and business and services

This requires employees with a background of business analysts to better understand the issues and understand the best way to communicate. In addition, there are differences between PPL and PPMN because each was intended for different purposes have strong features and conditions, and conditions, should be accessible to any type of cloud business or other cloud technology [33]. This applies to any type of project, as well as any terms, conditions, resources and submissions. Before choosing the best option, there are many tips and tricks to learn. Etro (2009) got off to a good start because user needs and problems can help determine which strategies to use [36]. A similar approach was adopted by Clems, Nimis and Sai (2008), who explain the importance of the business of the cloud and what communication requires. In all of this, Table 3 shows a list of tested methods [37].

8. Conclusion

This document describes the progress leading to the CCPF, as well as the use of cloud computing as a framework for effective work across all organizations [39]. It includes explains the various components with the CCPF works. These 4 framework show that the model is strategic and operates at the highest level of cloud performance for a top-down approach to defining needs and present a technical guide [40]. Content is influence by the business model and focuses on service delivery where revenue / profit is paramount to businesses.

References

[1] R. Rezaei, T. K. Chiew, S. P. Lee, no-Z. Shams Aliee, “A semantic integration framework for software as service programs in cloud computing environments,” Expert Systems with Applications, vol. 41, no. 13, pages 5751-5770, 2014.

[2] W. L. Shiau and P. P. Y. K. Chau, "Understanding the Behavioral Purpose of the Computer Computer Class:A Comprehensive Comparison Method," Information Management, vol. 53, pages 355-365, 2016.

[3] T. Dillon, C. Wu, noE. Chang, “Cloud computing: issues and challenges,” in Proceedings of the 24th IEEE International Conference on Advanced Information Networking and Applications (AINA '10), pages 27 –33, Perth, Australia, April 2010.

[4] M. Fan, S. Kumar, and A. B. Whinston, “Short-term and long-term competition between mobile software providers and software as a service,” European Journal of Operational Research, vol. 196, no. 2, pages 661-671, 2009.

[5] N. A. Sultan, "Access to the" cloud ": How to manage SMEs," International Journal of Information Management, vol. 31, no. 3, pages 272-278, 2011.
[6] M. Armbrust, A. Fox, R. Griffith et al., "The concept of cloud computing," Communications of the ACM, vol. 53, no. 4, pages 50-58, 2010.
[7] S. U. R. Malik, S. U. Khan, S. J. Ewen et al., "Performance analysis of deep cloud systems based on data management and replication: research," Distributed and Parallel Databases, vol. 34, no. 2, pages 179–215, 2016.
[8] S. Haag no-A. Eckhardt, "The adoption of the corporate cloud service: scientific analysis of literature and content," Journal of Business Economics, vol. 84, no. 3, pages 407-440, 2014.
[9] A. Benlian, M. Koufaris, and T. Hess, "The quality of service in software-as-a-service: Improving the SaaS-Qual standard and evaluating its role in continuous use," Journal of Management Information Systems, vol. 28, no. 3, pages 85-126, 2011.
[10] R. Buyya, CS Yeo, S. Venugopal, J. Broberg, et al. Brandic, "Cloud computing and emerging IT platforms: vision, hype, and reality for delivering computing as the 5th utility," Future Generation Computer Systems, vol. 25, no. 6, pages 599-616, 2009.
[11] T. Oliveira, M. Thomas, and M. Espadanal, "Exploring the frameworks for cloud computing: analysis of manufacturing and service sectors," Information and Management, vol. 51, no. 5, pages 497-510, 2014.
[12] V. Ratten, "The purpose of using cloud computing is continued: Innovativeness and creatives Perspectives," Journal of Business Research, vol. 69, no. 5, pages 1737-1740, 2016.
[13] UG. Garrison, S. Kim, noR. L. Wakefield, "Successful aspects of using cloud computing," Communications of the ACM, vol. 55, no. 9, pages 62-68, 2012.
[14] A. Duncan, S. Creese, and M. Goldsmith, "View all about internal computer attacks," Concurrency Computation, vol. 27, no. 12, pages 2964–2981, 2015.
[15] M. Ali, S. U. Khan, and A. V. Vasilakos, "Security in cloud computing: opportunities and challenges," Information Sciences. International Journal, vol. 305, pages 357-383, 2015.
[16] J. M. Del Alamo, R. Trapero, Y. S. Martin, J. C. Yelmo, N. Suri, "Exploring the privacy capabilities of cloud service providers," IEEE Latin America Transactions, vol. 13, no. 11, pages 3634-3641, 2015.
[17] UC. W. Autry, S. J. Grawe, P. J. Daugherty, noR. G. Richey, "The effects of the technological crisis and scope on the adoption and adoption of commercial technology," Journal of Operations Management, vol. 28, no. 6, pages 522-536, 2010.
[18] N. Phaphoom, X. Wang, S. Samuel, S. Helmer, and P. Abra-hamsson, "Studies conducted on major technological barriers affecting the decision to adopt cloud services," General of Systems and Software, vol. . 103, pages 167-181, 2015.
[19] A. Abdollahzadehgan, M. M. Gohary, A. R. C. Hussin, and M. Amini, "Organizational Criteria for Using Cloud Computing in SMEs," Journal of Information Systems Research and Innovation, vol. 4, no. 1, pages 67-74, 2013.
[20] T. Brandt, Y. Tian, M. Hedwig, noD. Neumann, "Autonomic management of Software as a Service systems with multiple quality service classes," at Proceedings of the 20 European Conference on Information Systems, (ECIS) '12), Barcelona, Spain, June 2012.
[21] J. Li, B. Li, T. Wo et al., "CyberGuarder: a framework for ensuring the security of green cloud computing," Future Generation Computer Systems, vol. 28, no. 2, pages 379-390, 2012.
[22] J. Spillner, G. Bombach, S. Matthischke, J. Mu¨ller, R. Tzschich- holz, and A. Schill, 4th IEEE / ACM International Conference on Cloud and Utility Computing, (UCC'11), pages 1–8, Melbourne, Australia, December 2011.
[23] Q.-A. Wang, C. Wang, K. Ren, W.-J. Lou, J. Li, "Enables public auditing and computer data security capabilities," Transaction IEEE on Parallel and Distributed Systems, vol. 22, no. 5, pages 847-859, 2011.

[24] C. Wang, S. S. Chow, Q. Wang, K. Ren, noW. Lou, "Privacy- which maintains public testing of secure cloud storage," developed by IEEE on Computers, vol. 62, no. 2, pages 362-375, 2013.
[25] UQ. Wang, C. Wang, K. Ren, W. Lou, andJ. Li, "Enables public auditing and data security storage for cloud computing," Transaction IEEEE on Parallel and Distributed Systems, vol. 22, no. 5, pages 847-859, 2011.
[26] G. Chen, Y. Wu, J. Liu, G. Yang, and W. Zheng, "Optimization of sub-query processing in distribution data data systems," Journal of Network and Computer Applications, vol . 34, no. 4, pages 1035-102, 2011.
[27] R. L. Grossman, Y. Gu, M. Sabala, noW. Zhang, "Compute and storage clouds using wide wide area performance," Future Generation Computer Systems, ivol. 25, no. 2, pages 179-183, 2009.
[28]M. Abu Sharkh, A. Kanso, A. Shami, noP. O¨ hle´n, "Building a Cloud in the World: A Study of Cloud Simulators in a Computing Data Center," Computer Networks, vol. 108, pages 78-96, 2016.
[29] A. Iosup, S. Ostermann, N. Yigitbasi, R. Prodan, T. Fahringer, noD. Epema, "The performance analysis of cloud computing services for multiple computer applications," IEEE Transaction on Parallel and Systems, vol. 22, no. 6, pages 931-945, 2011.
[30] A. Lin and N.-C. Chen, "Cloud computing as an innovation: Percepetion, attitude, and adoption," International Journal of Information Management, ivol. 32, no. 6, pages 533-540, 2012.
[31] G. Feuerlicht and S. Govardhan, "Impact of cloud computing: beyond a technology trend," in Proceedings of The International Conference on Systems Integration, Oeconomica, vol. 1, page. 8, Prague, Czech Republic, 2010.
[32] G. Feuerlicht, L. Burkon, and M. Sebesta, "Cloud computing adoption: what are the issues?" System Integration, vol. 18, no. 2, pages 187-192, 2011.
[33] P. Ge´czy, N. Izumi, and H. Koˆiti, "Cloudsourcing: cloud hosting management," Global Journal of Business Research, vol. 6, no. 2, pages 57-70, 2012.
[34] T. Ackermann, A. Benlian, T. Widjaja, and P. Buxmann, "It has been found that IT security threats use cloud computing: thinking and scale development," in Proceedings of the International Conference on Information Systems , (ICIS '12), pages 3468-3487, Orlando, Fla, USA, December 2012.
[35] A. Benlian and T. Hess, "Software-as-a-Service Opportunities and Dangers: Findings in IT Management Research," Decision Support Systems, vol. 52, no. 1, pages 232-246, 2011.
[36] B. Martens and F. Teuteberg, "Risk and control of cloud computing services: the design of a reference model," in Proceedings of the 17th Americas Conference on Information Systems 2011, (AMCIS'11), pages 2041. -2050, Detroit, Mich, USA, August 2011.
[36] D. Yuan, Y. Yang, X. Liu, J. Chen, "The estimated low cost of stock in the middle of a cloud workflow system," Journal of Parallel and Distributed Computing, vol. 71, no. 2, pages 316-332, 2011.
[37] W. Pauley, "transparency transparency: Monitoring intensive," IEEE Security and Privacy, ivol. 8, no. 6, pages 32-39, 2010.
[38] Y.-M. Wang, Y.-S. Wang, and Y.-F. Yang, "Understanding RFID Acquisition Decisions in the Manufacturing Industry," Technology Forecasting and Social Change, vol. 77, no. 5, pages 803-815, 2010.
[39] I. Ion, N. Sachdeva, P. Kumaraguru, and S. Ćrapkun, "Home is safer than the cloud! The privacy concerns of consumer cloud storage, "in Proceedings of the 7th Symposium on Usable Privacy and Security, (SOUPS'11), Pittsburgh, Pa, USA, July 2011.
[40] S. Pearson, "Toward accountability in the cloud," IEEE Internet Computing, p. 15, no. 4, pages 64-69, 2011.

Effect of variation in channel length and thickness of organic semiconductor in the bottom-gate configuration of OTFT

S Gupta[1], **M K Singh**[2]

[1,2]Department of Electronics Engineering, Harcourt Butler Technical University (Formerly HBTI) Kanpur, UP, INDIA-208002

[2]E-mail: mksingh@hbtu.ac.in

Abstract-Advancing in the device and circuit performance a vast amount of study is going in the area of organic electronics. In the present work, we have patterned the features of the bottom-gate configuration of OTFT with the help of Technology Computer-Aided Design (TCAD) tools where ATLAS^{TM} (Silvaco International) has been utilized as a 2D numerical device simulator to extract the electrical characteristics. To monitor the effect of variation in channel length and thickness of OTFT, we have altered the channel length (L) from 10um to 50um, which results in the reduction of the drive current from -1.4×10^{-10} A to -2.7×10^{-11} A (BGBC), and in contrary to that the performance of the device has been affected as the current ratio decrease from 1.2×10^{12} to 1.4×10^{3} (BGBC) with a thickness variation of organic semiconductor from 10nm to 50nm. Furthermore, the electrical characteristics as well as performance parameters like mobility, on/off current ratio, threshold voltage, etc., related to configured Bottom-Gate (top and bottom contacts) pentacene-based OTFT devices have been retrieved and compared.

Keyword- Organic thin film transistor, bottom-gate configurations, Organic semiconductor thickness, channellength variation, numerical simulation, electrical parameter.

1. Introduction

Organic electronics is emerging as an alternative to change conventional electronics' modernization due to unique features like flexibility, low cost, ease of processing, ultrathin, etc. The backbone of the organic electronics is the conducting organic material, which was firstly reported by Heeger, Diarmid, and Skirakawa in 1977. For this novel invention, they were combinedly awarded the noble prize in 2000 [1]. After the influx of conducting organic material, organic devices came into the picture. The first organic field-effect transistor in 1986 was based on an organic macromolecule, polythiophene, used as the semiconductor (active layer) [2], which assisted in having better charge transport. Simultaneously, organic material mobility was observed to be less in contrast to the traditional semiconductor material. Research is going on to increase the mobility of organic field-effect transistor (OFET) devices by using different high performances of organic material and fabrication techniques to help these devices to be a part of plastic electronics, textiles electronics, and robotic skin [3-7]. All these implementations are only possible with the help of a transistor called an organic field-effect transistor. In recent times these pentacene organic field-effect transistors demonstrate in "functional logic circuitry[8] and active-matrix liquid crystal display [9]. In the current scenario, the devices

related to organic electronics can compete against conventional electronics in the marketplace due to their novel features.
The working principle of organic thin-film transistor (OTFT) is like a metal oxide semiconductor field-effect transistor (MOSFET). Still, the channel formation differs, and it works in the voltage-controlled current source when we apply a voltage between gate and source (V_{gs}), the accumulation process of the charge carrier starts at the interface of semiconductor and insulator, which help in the generation of current and voltage applied between drain and source (V_{ds}) help in the flow of current. OTFT is incorporated with the organic semiconductor layer, gate dielectric, and the electrodes, which include gate, drain, and source. These layers are arranged in different devices structure mainly; we determine the structure of OTFT by the position of the gate dielectric, which can be upward and downward. As a result, we termed as the top gate (TG) and bottom gate (BG) configuration and according to the source's location and drain contact regarding the organic semiconductor layer, which is also divided into top and bottom contact structure.

This paper has discussed the OTFT with bottom gate configuration, where we have simulated the 2D structure and shown the comparative study between bottom gate top contact (BGTC) and bottom gate bottom contact (BGBC) structured OTFTs with high dielectric constant. Figure 1 shows the structure of the top and bottom contact bottom gate OTFT devices.The scope of application of these OTFTs is decided by various parameters, which mainly include mobility, threshold voltage, on/off current ratio and sub-threshold slope, etc. [10-12]; these parameters have been extracted using device simulator software ATLAS from Silvaco International [13]. Where the prime variations are being done in structural dimension, the material used in layers, the grain size of organic semiconductor thin film, and morphology of the semiconductor. Here also we have changed the structural dimension parameter like channel length [14] and thickness of the organic semiconductor, i.e., the polymer used [15], which affects the performance of the OTFT. Due to good organic material's conductivity and flexibility, it replicates analogous operations like other conventional electronics devices such as LED, TFT, FET, and many more.

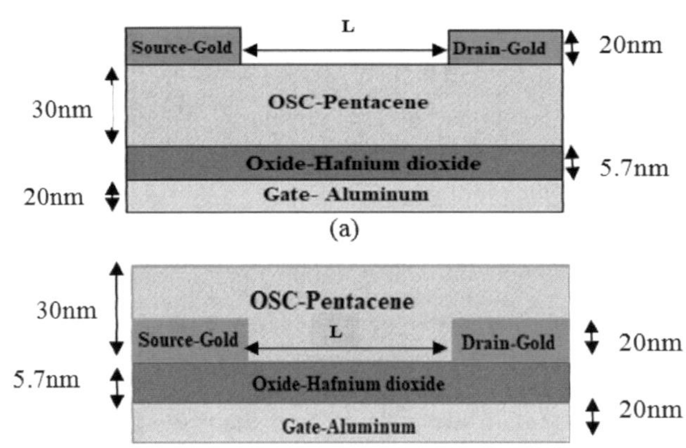

Figure 1. Bottom gate OTFT (a) Bottom gate top contact (b) Bottom gate bottom contact

2. Device structure simulation and modeling setup

We have modeled the bottom gate configurations with top and bottom contact structure using an analytical model based simulator to realize the device physics. Examining the electrical characteristics by designing the top and bottom contact OTFTs, which is formed with the help of the Silvaco Atlas 2-D numerical device simulator. For comparison of top and bottom contact structure, an Aluminum gate electrode with thickness 0.02 μm is selected. HfO_2 gate dielectric with high dielectric constant 22.0 is used and deposited above gate electrode with thickness 5.7nm. The pentacene layer that is p-type organic semiconductor material is deposited over the oxide layer with a thickness of 30nm with a doping concentration of 3×10^{17}, generation of the conducting channel took place.Since this is a simulated work but during the practical device fabrication, similar steps are involved as proposed by many researchers [16-23]. Depostion of the materials plays a central activity while determining the performance of different OTFTs. The generally used method for the deposition of thin-film of polymers on the substrate is a solution-processed spin-coating method [18, 19].But the novel alternative approach for well-oriented, uniformly deposited polymer thin-film (OSC) is proposed as floating film transfer method (FTM) which has been widely adopted during the fabrication of PQT-12 polymer based practical OTFTs [20-23]. Further, for the metal contacts (Au) have been deposited over as deposited (source and drain) over active layer of OSC.In the present work the deposition of the source and the drain electrode is changed according to the structure with a thickness of 20 nm. Moreover, In table 1. the structure-based dimensional parameter with material and thickness is summarized [24,25]. The simulated device structure in TCAD Silvaco is shown in fig. 2.

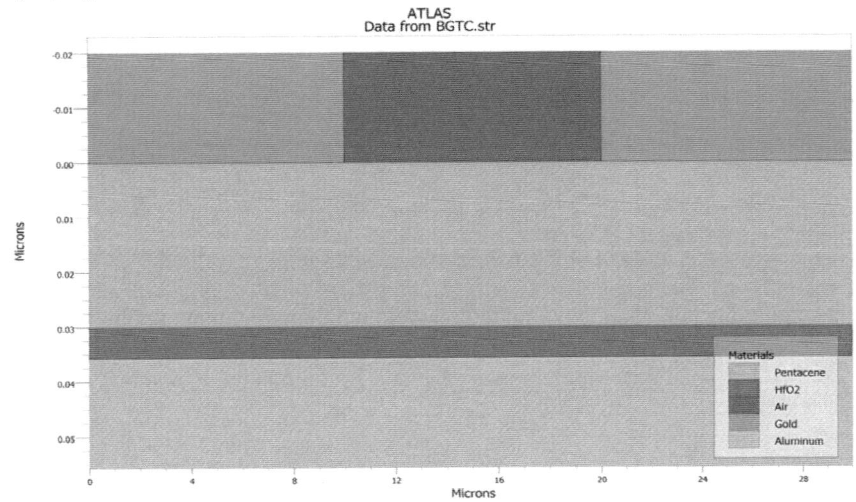

(a)

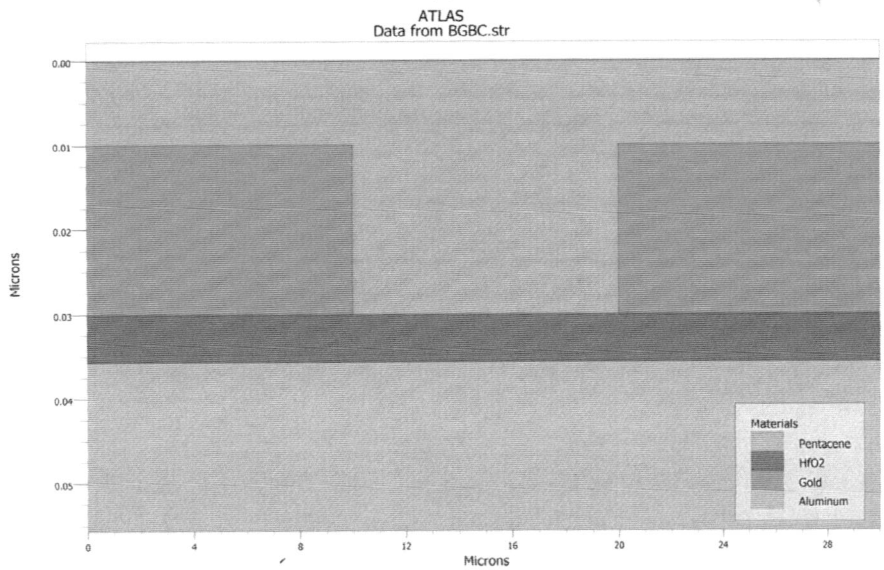

(b)

Figure 2. TCAD Silvaco Simulated Bottom gate configurations of (a) Top contact and (b) Bottom contact

The procedure comprises mathematical models such as Poisson's, Continuity, and transport equation that assist the TCAD tool in displaying the electrical parameter which is associated with the performance of the device.

2.1. Physical equation of devices:

Poisson's Equation associates between electrostatic potential and local charge densities. Equation 1 and 2 expresses the mathematical relationship.

$$\nabla.(\varepsilon\nabla\psi) = -\rho \qquad (1)$$

$$\nabla.(\varepsilon\nabla\psi) = -q(p - n + N_d^+ - N_a^-) \qquad (2)$$

Where ψ and ε represent electrostatic potential, and local permittivity and ρ represents local space charge density.During the simulation, it always takes the intrinsic Fermi potential as a reference potential. The sum of fixed and mobile charges plays a significant role in the local space charge density. The sum includes the electron density (n), hole density (p), and ionized impurities such as N_d^+ represents ionized donor density, and N_a^- represents ionized acceptor density.

Continuity equation represents the dynamic of charge carrier arrangement with time. The equation for electron and holes as shown in equation 3 and 4:

$$\frac{\partial n}{\partial t} = \frac{1}{q}\nabla.J_n + G_n - R_n \qquad (3)$$

$$\frac{\partial p}{\partial t} = \frac{1}{q}\nabla.J_p + G_p - R_p \qquad (4)$$

Herein the above equations, the electronic charge magnitude has been represented by q , n and p are the electron and hole concentration, G and R are the corresponding generations and recombination rate. J is the current density. During the simulation, both Equations 3 and 4 take part, but we can turn off a particular equation that is not used for a specific model.

The transport equation, specify the device's physics for the electron and hole current densities, generation, and recombination rates. It uses a drift-diffusion charge transport model for the current density equation. Equation 5 and 6 describe the drift-diffusion model as follows:

$$J_n = qn\mu_n E_n + qD_n\nabla_n \qquad (5)$$

$$J_p = qn\mu_p E_p + qD_p\nabla_p \qquad (6)$$

Where n and p are the electron and hole density, μ is the carrier mobility, E is the local electric field, D is the diffusion coefficient. Equation 7 and 8 describe the local electric field as follows:

$$E_n = -\nabla(\psi + \frac{kT_L}{q}\ln n_{ie}) \qquad (7)$$

$$E_p = -\nabla(\psi + \frac{kT_L}{q}\ln n_{ie}) \qquad (8)$$

Wheren_{ie} and T_L represent the local effective intrinsic concentration and lattice temperature, respectively.

The Poole-Frenkel mobility model has been used to investigate the device's static and dynamic operation with finite element-based simulation. It can be expressed as-

$$\mu(E) = \mu_0 \exp\left[-\frac{\Delta}{KT} + \left(\frac{\beta}{KT} - \gamma\right)\sqrt{E}\right] \qquad (9)$$

Here, $\mu(E)$ shows field-dependent mobility, μ_0 is zero-field mobility which is depicted when the drain voltage is zero, E is the electric field and Δ, β, γ is the activation energy, hole Poole-Frenkel constant, fitting parameter Table 2[24]. Shows the input parameter of the bottom-gate OTFT used during the simulation.

2.2. Density of defect states model

The total density of defect states (DOS), g(E), comprises four bands, two tail and two deep state function such as an acceptor-like ($g_{TA}(E)$)and a donor –like($g_{TD}(E)$) exponential band tail function, an acceptor-like ($g_{GA}(E)$) and a donor- like ($g_{GD}(E)$) Gaussian deep state function. The equation 10 specified as total DOS:

$$g(E)= g_{TA}(E)+ g_{TD}(E)+ g_{GA}(E)+ g_{GD}(E) \quad (10)$$

Where, E specify the state energy, subscripts (T, G, A, D) is tail, Gaussian (deep level), acceptor, and donor states. The equation 11 to 14 describing the term used in equation 10.

$$g_{TA}(E) = N_{TA}\exp[\frac{E-E_C}{W_{TA}}] \quad (11)$$

$$g_{TD}(E) = N_{TD}\exp[\frac{E_V-E}{W_{TD}}] \quad (12)$$

$$g_{GA}(E) = N_{GA}\exp[-[\frac{E_{GA}-E}{W_{GA}}]^2] \quad (13)$$

$$g_{GD}(E) = N_{GD}\exp[-[\frac{E-E_{GD}}{W_{GD}}]^2] \quad (14)$$

In exponential tails, DOS is specified by conduction and valence band intercept densities (N_{TA} and N_{TD}) and its characteristic decay energy (W_{TA} and W_{TD}). In Gaussian distributions, DOS is specified by the total state density ($N_{GD}$and N_{GA}), and its characteristic decay energy ($W_{GD}$and W_{GA}), and peak energy distribution (E_{GA} and E_{GD}).

Table 1.Device dimension of bottom-gate configuration top and bottom contact with the material

Device Dimension	Material	Value
The thickness of OSC(t_{OSC})	Pentacene	30nm
The thickness of dielectric(t_{OX})	HfO_2	5.7nm
The thickness of the gate electrode (t_G)	Aluminum	20nm
Thickness of S/D contact($t_{S/D}$)	Gold	20nm
Channel length	-	10um
Channel width	-	220um

Table 2. Simulation input parameter of OTFT

Parameters	Values
The work function of source & drain(Au)	5.1
The work function of Gate (Al)	4.28
Doping Concentration	3×10^{17} cm-2
The dielectric constant for HfO2	22
Permittivity for pentacene	4.0
The activation energy for zero electric fields for holes	1.792×10^{-2} eV
Hole Poole-Frenkel factor	7.758×10^{-5}eV $(cm/v)^{1/2}$
Effective density of state in the conduction band (Nc)	1.0×10^{21} cm-3
Effective density of state in the valance band (NV)	1.0×10^{21} cm-3

3. Results and discussions

The electrical characteristics of bottom-gate configurations of top and bottom contact, electrical parameters like threshold voltage, subthreshold voltage, on/off ratio, transconductance, and mobility are evaluated and compared with the help of numerical simulation.

3.1. Comparision and performance of the bottom-gate top and bottom contact OTFT

The electrical characteristics of the top and bottom gate configuration are being compared. After the simulation by using the above dimensional parameter, we show the outcomes in Table 3. we analyzed that the on/off current ratio and mobility is high of bottom contact than top contact. It is observed bottom gate Bottom contact shows the superior result when compared with top contact at gate voltage greater than 7.5 V with drain voltage -1V because the traveling path of charge carrier between source and drain is unlike [26] shown in figure 3 transfer characteristics curve.The dielectric material modification leads an essential part in determining the electrical parameter. Here, increase in the capacitance, C by a factor of k (dielectric constant) as shown in equation 15

$$C = \frac{\varepsilon_0 k}{t_{ox}} \quad (15)$$

Where t_{ox} is the dielectric layer thickness. The increased capacitance leads to a shift in charge distribution in dielectric material because of polarization induced by an external electric field. The high-k dielectric material is helpful for the low threshold voltage. Similarly, high on-current (with high current ratio) and a more downward subthreshold slope are observed [10]. Table 4 shows the comparison of the electrical parameter according to the reported literature.Here, low mobility results from surface roughness directing to the morphological disorders in OSC, which reduces the carrier's mobility.

Table 3.Electrical parameter extraction of BGTC and BGBC

Parameters	BGTC	BGBC
Threshold voltage V_T (V)	-0.601	-0.666
Sub-Threshold Slope SS (V/Decade)	0.291	0.284
Current On-Off Ratio I_{ON}/I_{OFF}	3.7×10^4	4.1×10^4
Transconductanceg_m(μS)	4.34×10^{-6}	4.96×10^{-6}
Mobility μ (cm^2/VS)	0.0578	0.0661

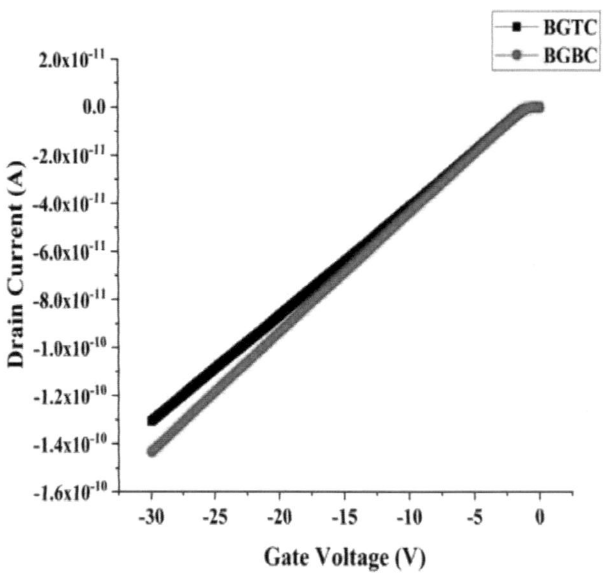

Figure 3. Bottom gate top and bottom contact transfer characteristic curve at drain voltage -1V

Table 4. Comparison of the electrical parameter of top and bottom contact bottom gate OTFT with the different dielectric material.

Material of layer	Structure	μ (cm^2/Vs)	$I_{ON/OFF}$	V_T(V)	SS (V/dec.)	Supply voltage V_{ds}(V)	V_{gs}(V)	Ref.
OSC-Pentacene, S/D-Gold, G- Aluminum, D-Aluminum oxide (Al_2O_3)	BGTC BGBC	NR NR	6.9×10^3 2.2×10^2	-0.5 -0.4	0.34 1.31	-1	0 to -20	24
OSC-Pentacene, S/D-Gold, G- Aluminum, D-Aluminum oxide (Al_2O_3)	BGTC BGBC	0.129 0.14	9.5×10^3 1.9×10^4	-1.2 -1.1	0.32 0.26	-1	0 to -30	26
OSC-Pentacene, S/D-Gold, G- n^+ Si, D-Silicon dioxide(SiO_2)	BGTC BGBC	0.085 0.0014	NR NR	-3.2 -8.5	NR NR	-25	0 to -20	27
OSC-Pentacene, S/D-Gold, G- Aluminum, D- Hafnium oxide (HfO_2)	BGTC BGBC	0.057 0.066	3.7×10^4 4.1×10^4	-0.60 -0.66	0.29 0.28	-1	0 to -30	Present Work

Where G and D represent the gate and dielectric layer, respectively, S/D is source/drain. $V_{ds}$is the drain to source voltage, and V_{gs} is the gate to source voltage. V_T and SS are the threshold voltage and subthreshold slope, μ is the mobility, and NR is not reported.

3.2. Effect of variation of channel length and organic semiconductor thickness on bottom gate top and bottom contact OTFT.

This segment represents the dimensional parameters influences, such as channel length from 10μm to 50μm and organic semiconductor thickness from 10nm to 50nm with a step size of 10. It shows the variation (in both cases) in the operation of bottom-gate OTFT significantly.

We show by increasing the channel length, the drive current reduces in Figures 4. (a) and (b). Such variation may be because channel length is inversely proportional to drive current [28]. We see mobility increment in BGBC while reducing the channel length. And other parameters such as the threshold voltage reduces, we see a slight variation in the current ratio($I_{ON/OFF}$) and also affect the other electrical parameter of devices, which is summarized in table 5.

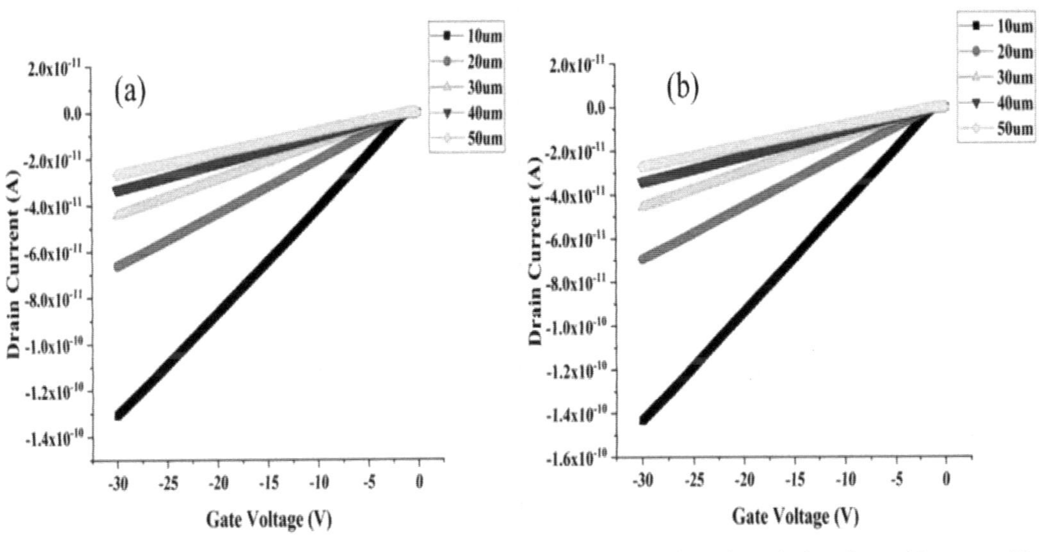

Figure 4. Drain current variation due to variation in the channel length variation from 10 μm to 50μm in (a) BGTC (b) BGBC

Table 5. Electrical parameter of bottom-gate top and bottom contact OTFT during channel length variation

Channel length	Device type	Threshold voltage V_T (V)	I_{ON}(A)	I_{OFF}(A)	$I_{ON/OFF}$	Sub-threshold Slope SS (V/decade)	Mobility μ(cm^2/VS)	Transconductances g_m (μS)
10μm	BGTC	-0.601	-1.309×10^{-10}	-3.426×10^{-15}	3.78×10^{4}	0.291	0.0578	4.34×10^{-6}
	BGBC	-0.666	-1.437×10^{-10}	-3.430×10^{-15}	4.16×10^{4}	0.284	0.0661	4.96×10^{-6}
20μm	BGTC	-0.615	-6.626×10^{-11}	-1.691×10^{-15}	3.91×10^{4}	0.297	0.0597	2.24×10^{-6}
	BGBC	-0.679	-6.945×10^{-11}	-1.687×10^{-15}	4.11×10^{4}	0.287	0.0639	2.40×10^{-6}
30μm	BGTC	-0.623	-4.427×10^{-11}	-1.119×10^{-15}	3.95×10^{4}	0.303	0.0603	1.51×10^{-6}
	BGBC	-0.685	-4.568×10^{-11}	-1.116×10^{-15}	4.09×10^{4}	0.287	0.0631	1.58×10^{-6}
40μm	BGTC	-0.628	-3.320×10^{-11}	-8.359×10^{-16}	3.97×10^{4}	0.310	0.0602	1.13×10^{-6}
	BGBC	-0.689	-3.399×10^{-11}	-8.334×10^{-16}	4.07×10^{4}	0.287	0.0622	1.17×10^{-6}
50μm	BGTC	-0.632	-2.654×10^{-11}	-6.671×10^{-16}	3.97×10^{4}	0.317	0.0608	9.13×10^{-7}
	BGBC	-0.691	-2.704×10^{-11}	-6.648×10^{-16}	4.06×10^{4}	0.287	0.0623	9.38×10^{-7}

Increment in organic semiconductor thickness, the behavior of devices affected by the bulk traps and charge localization which is found in the interface of OSC-dielectric because of which the devices performance changes. The ON current of the device is affected by means of the variation in thickness of the semiconductor. However, an increase in OFF current with an increase in semiconductor (OSC) thickness is due to the subsequent rise in the bulk current. The switching activity drop-off with the larger magnitude of OFF current, which increases the power consumption and charge leakage in the device. Figure 5 shows transfer characteristics of both structures at V_{ds}=-1. Subsequently, decreasing the OSC thickness in BGTC, an increase in drain current (slight variation in case of BGBC based device) has been observed. From the current density profile, a very nominal variation in the drain current has been analyzed. The threshold voltage is reduced because of an increment in the free charge carrier with an increase in active layer thickness. Mobility in BGTC increases on increasing the thickness of the OSC layer. This is due to the decrease in the average value of the charge carriers [29,30]. From table 6, the same can be observed clearly along with other electrical parameters.

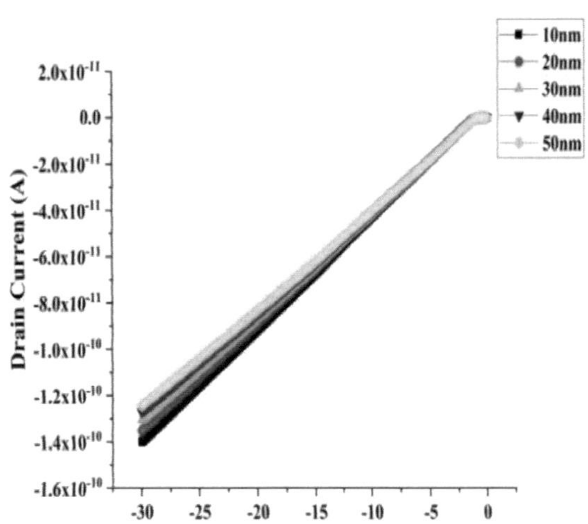

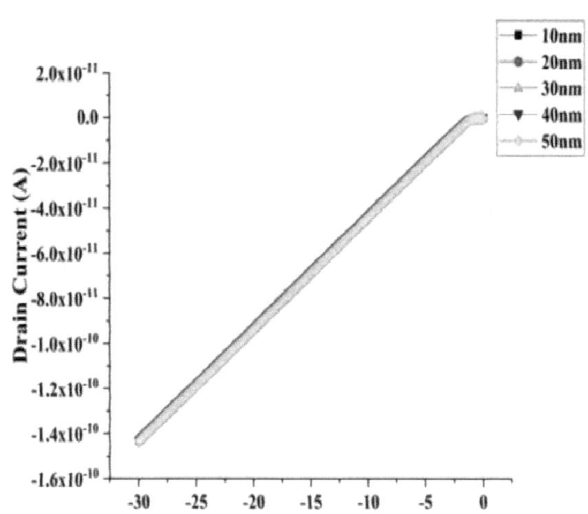

(a)

(b)

Figure 5. Drain current variation due to variation in the organic semiconductor thickness variation from 10 nm to 50 nm in (a) BGTC (b) BGBC

Table 6. Electrical parameter of bottom-gate top and bottom contact OTFT during organic semiconductor thickness variation

OSC Thickness	Device type	Threshold voltage V_T (V)	I_{ON}(A)	I_{OFF}(A)	$I_{ON/OFF}$	Sub-threshold Slope SS (V/decade)	Mobility μ(cm^2/VS)	Transconductances g_m (μS)
10*nm*	BGTC	-0.663	-1.401×10^{-10}	-5.301×10^{-19}	2.6×10^{8}	0.065	0.0640	4.82×10^{-6}
	BGBC	-0.695	-1.431×10^{-10}	-1.133×10^{-22}	1.2×10^{12}	0.036	0.0659	4.96×10^{-6}
20*nm*	BGTC	-0.626	-1.350×10^{-10}	-8.072×10^{-19}	1.6×10^{8}	0.036	0.0606	4.56×10^{-6}
	BGBC	-0.681	-1.436×10^{-10}	-5.447×10^{-19}	2.6×10^{6}	0.060	0.0662	4.98×10^{-6}
30*nm*	BGTC	-0.601	-1.309×10^{-10}	-3.462×10^{-15}	3.7×10^{4}	0.291	0.0578	4.34×10^{-6}
	BGBC	-0.666	-1.437×10^{-10}	-3.430×10^{-15}	4.1×10^{4}	0.284	0.0659	4.96×10^{-6}
40*nm*	BGTC	-0.579	-1.271×10^{-10}	-3.256×10^{-14}	3.9×10^{3}	0.636	0.0554	4.17×10^{-6}
	BGBC	-0.652	-1.434×10^{-10}	-3.261×10^{-14}	4.3×10^{3}	0.633	0.0659	4.96×10^{-6}
50*nm*	BGTC	-0.560	-1.244×10^{-10}	-9.588×10^{-14}	1.2×10^{3}	0.765	0.0535	4.03×10^{-6}
	BGBC	-0.638	-1.434×10^{-10}	-9.600×10^{-14}	1.4×10^{3}	0.761	0.0659	4.96×10^{-6}

4. Conclusion

The performances of the bottom-gate configured top, as well as bottom contact OTFTs based on pentacene with high dielectric constant material, have been analyzed and compared. It is observed that BGBC shows a much better result as compared to BGTC. Besides this, the effect of channel length and organic semiconductor thickness is examined. We observed lower channel length produces a high drive current, and decreasing organic semiconductor thickness reduces the leakage current and leads to an increment in the current ratio. Such results are much desirous in displays and memory circuits.

Acknowledgment

Dr. M. K. Singh is thankful to Prof. Rajiv Prakash, IIT (BHU) Varanasi, for his fruitful suggestions and support to access the TCAD. The authors are grateful to all faculty members & Head, Department of Electronics Engineering, HBTU Kanpur, for giving time to time help and support. S. Gupta is cordially thankful to the MHRD Govt. of India/HBTUKanpur for financial assistance through the GATE fellowship during her M.Tech course.

References

[1] Shirakawa H, Louis E J, Macdiarmid A G, Chiang C K and Heeger A J 1977*J. Chem. Soc. Chem. Commun.***39**, pp. 1098-1101.
[2] Tsumura A, Koezuka H and Ando T 1986 *Appl. Phys. Lett.* **49**, pp. 1210-12
[3] Klauk H, Gundlach D J and Jackson T N 1999 *IEEE Electron Device Lett.* **20**, pp. 289-291.
[4] Halik M, Klauk H, Zschieschang U, Schmid G, Radlik W, Ponomarenko S, kirchmeyer S and Weber W 2003 *Journal of Applied Physics***93**, pp. 2977-81.
[5] Crone B K, Dodabalapur A, Sarpreshkar R, Filas R W, Lin Y-Y, Bao Z, O'Neil J H, Li W and Katz H E 2001 *J. Appl. Phys.***89**, pp. 5125-32.
[6] Lee J B and Subramanian V 2003*IEEE International Electron Devices Meeting (IEDM'03)* pp. 199-202.
[7] Someya T and Sakurai T 2003*IEEE International Electron Devices Meeting* pp. 203-6.
[8] Kane M G, Campi J, Hammond M S, Cuomo F P, Greening B, Sheraw C D, Nichols J A, Gundlach D J, Huang J R, Kuo C C, Jia L, Klauk H and Jackson T N 2000 *IEEE Electron Device Letter***21**, pp. 534-6.
[9] Sun H, Yin Y, Wang Q, Jun Q, Wang Y, Tsukagoshi K, Wang X, Hu Z, Pan L, Zheng Y, Shi Y and Li Y 2015 *Appl. Phys. Lett.* **88**, pp. 053304-1-4.
[10] Kumar B, Kaushik B K and Negi Y S 2014 *Polym. Rev.* **54**, pp. 33-111.
[11] Es-Saghiri A, Boufounas E-M, El Amrani A and Lucas B 2018*4th International Conference on Optimization and Applications(ICOA).*

[12] Sharma S and Varma T 2018 *Materials Research Express***6**, pp. 1-11.
[13] ATLAS User's Manual from SILVACO International.
[14] Mittal P, Kumar B, Negi Y S, Kaushik B K and Singh R K 2012 *Microelectronics Journal***43**, pp.985-994.
[15] Resendiz L, Estrada M, Cerdeira A, Iniguez B and Deen M J 2010*Org. Electron***11**, pp. 1920-27.
[16] Singh S, Tiwari P K, Kumar H, Kumar Y, Rawat G, Kumar S, Singh K, Goel E, Jit S, Park S 2017 *NANO*, pp. 1750137-1-8
[17] Singh MK, Pandey R K, Prakash R2017 *Organic Electronics* 2017 **50**, pp. 359-66.
[18] Singh M K, Kumar A, Prakash R 2017*Materials Science and Engineering:B***217**, pp. 12-7.
[19] Singh M K, Kumar A, Prakash R 2018*Organic Electronics***54** pp. 209-15.
[20] Kumar C, Rawat G, Kumar H, Kumar Y, Prakash R, Jit S 2018 *IEEE Transactions on Nanotechnology***17**, pp. 1111-7
[21] Kumar C, Rawat G, Kumar H, Kumar Y, Prakash R, Jit S 2017 *12th IEEE Nanotechnology Material and Devices Conferences* pp. 37-38
[22] Kumar C, Rawat G, Kumar H, Kumar Y, Kumar A, Prakash R, Jit S 2018 *IEEE Sensors Journal***18**, pp. 6085-91
[23]Kumar C, Rawat G, Kumar H, Kumar Y, Prakash R, Jit S 2017 *Organic Electronic***48,** pp. 53-68
[24] Singh A and Singh M K 2020 *Asian Journal of Convergence in Technology***6**, pp. 1-4.
[25] Singh A and Singh M K 2020 *IEEE International Conference for Innovation in Technology*pp. 1-5.
[26]Kumari P and Dwivedi A D D 2019*Global Journal of Research In Engineering***19**, pp. 6-12.
[27]Gupta D, Katiyar M, Gupta D 2009 *Organic Electronic***10**, pp. 775-784
[28]Singh A and Singh M K 2020 *International Conference on Advances in Computing, Communication & Materials (ICACCM)* pp. 301-5.
[29] Kumar B, Kaushik B K, Negi Y S 2014 IET Circuits, Devices & Systems **8,** pp. 131-140
[30] Singh A and Singh M K 2020 *Journal of Physics: Conference Series***1706**, pp. 1-12

IOP Conf. Series: Materials Science and Engineering 1119 (2021) 012013 doi:10.1088/1757-899X/1119/1/012013

Analysing Uber Trips using PySpark

Rahul Pradhan[1], Praveen Kumar Mannepallli[2] and Vikram Rajpoot[3]

[1,3] Department of CEA, GLA University, Mathura

[2] School of Computer Science and Technology, LNCT University, Bhopal

E-mail: [1]rahul.pradhan@gla.ac.in, [2]praveen.hawassa@gmail.com, [3]vikram.rajpoot@gla.ac.in

Abstract. All the enterprise has a lot of data. To develop business, on occasion records evaluation required. By examining facts, we get vital matters on which work out and make our graph for the future via which made best future decisions. Most of the organizations going on line the place the statistics generate will increase day by using day. To develop commercial enterprise with this aggressive surroundings records evaluation is necessary. This analytics venture is very important to recognize the use of records analytics. Through initiatives like this, many organizations can recognize a number of complicated operations. Uber Data Analysis task permits us to recognize the complicated facts visualization of this large organization. It is developed with the assist of 'R' programming language. In this venture we analyze the Daily, Monthly and Yearly Uber Pickups in New York City. This mission is primarily based on Data Visualization that will information you toward use of ggplot2 library for perception the data.

1. Introduction

This paper is based totally on facts storytelling which is an essential element of Machine Learning through which organizations are capable to apprehend the historical past of quite a number operation. With the assist of visualization, businesses can avail the advantage of appreciation the complicated information and acquire insights that would assist them to craft decisions. This assignment is associated to large facts as we are inspecting very big quantity of records to be aware of the instinct of uber customers. This challenge will exhibit the plotting of daily, month-to-month and every year rides of uber in a complete city. On-demand, app-based trip offerings like Uber and Ola have end up an necessary section of today's transportation gadget with its flexibility and speedy responsiveness. Compared with regular taxicabs, Uber-like taxis have loggers to reveal and document time out statistics such as pickup place and outing distance, which can be a precious facts supply for information discovering. Nowadays, a Real-time prediction for experience provider demand (always mirrored by way of the wide variety of pickups) is increasingly more critical for the motive of enhancing the effectively and sustainability of the city transportation system. Newly aroused utility matters like experience sharing and independent mobility dispatching are primarily based on strong demand predictions. In this paper, the authors advocate a deep gaining knowledge of primarily based strategy to make dynamic predictions for uber pickups the use of historical data [1].

1.1. Predicting the number of uber pickups

The dataset we used for this challenge blanketed records of Uber cars' ridership in the metropolis of New York for the six months of 2014. we was once capable to model the demand with

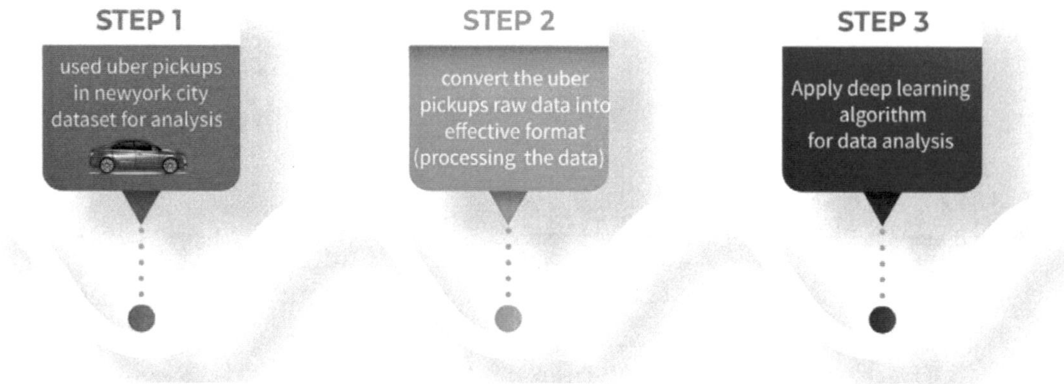

Figure 1. Three step process model for uber data analysis

forecasting horizons from a week to subsequent hour. These models can be used in extraordinary occasions. For instance any person may want to use the weekly forecasting model to have established view of the subsequent week's demand. On the different hand, a actual time gadget may want to examine the prediction per borough, with the positions of Uber automobiles and spotlight the areas hence to drivers' functions assisting them to roam greater correctly through the city. Since the model is based totally on previous observations it is inclined to incorrect estimations on very irregular conditions. Additionally considering modern observations have an effect on future prediction, demand out of the normal degrees may also lead to incorrect estimation at some factor to later predictions [2]. (see figure 1)

2. Related Work

Initial research about the demand for digital mail services is developing rapidly, To find out about the forecast overall performance of the research models, you pick to select information for peculiar day. The Uber information incorporates records about the role and time of the trips and returns of every day trip in the course of a day. According to the reachable dataset, the Uber trips historic information of Apr. 2014. This research proposes a strategy for examining and predicting the Uber taxi demand [3]. This lookup research papers the first-class of travel time in the city of New Delhi. They acquire information on 610 journeys from 34 customers the usage of the user's cell and net applications. They empirically exhibit the unpredictability of tour time estimates for first-rate taxis. This is the poor consequences of this unpredictability on passengers ready for taxis, which leads to the cancellation of a big 28.4 percentage of them. The empirical observations vary appreciably from the excessive accuracies mentioned in journey time estimation literature.(a) statistics trouble for growing international locations or (b) that can't seize the historic patterns in growing vicinity tour instances or (c) a mindful policy choice by way of Uber platform or Uber drivers [4]. Initial lookup has proven that through the use of designated facts on taxis at the journey degree and on the rental car and statistics on complaints about the degree of new complaints at the degree of incidents, we learn about how Uber and Lyft enter broken the first-rate of taxi offerings in New York City. The usual impact of the companies primarily based on the state of affairs and in unique of the using administrations was once big and widespread. One of these results is the enlargement of the competition between Uber and Lyft over the fantastic of taxi administration. They use a new set of grievance information to measure (the lack of) high-quality of carrier that we have in no way been analyzed before. Focus on the pleasant dimensions generated through most of the complaints we demonstrate.

The multiplied opposition for these shared journey offerings has had an intuitive have an impact on the conduct of taxi drivers [5]. It explains a massive quantity of space-time facts is generated via millions of buses in metropolitan cities round the world. These dataset, if analyzed correctly, can furnish a higher appreciation of the demand for taxis. With the developing choice of clients to have a easy experience, thanks to their point-to-point service, taxis are turning into the first-rate choice for everyone. This record analyzes Uber's journeys to New York from April 2014 to September 2014 for the evaluation and detection of essential points. The main facts bought have been labeled into three principal categories: morning of the week, afternoon of the week and top weekend hours. For the detection of crucial points. Spatial records in the shape of longitude and latitude are used for geographical mapping of withdrawal positions primarily based on the day of the week and the time of day. Clustering methods based totally on medium PAM k have been applied in the spital information to decide the pleasant strategy for the detection of strong and efficient fundamental factors [6].

3. Proposed work and analysis

We use NYC dataset contain data on over 4.5 million trips in NYC from Apr. to Sept. 2014. There are the files of dataset on Uber trips in NYC from Apr. to Sept. 2014 [7]. The files are separated by month and the following columns:

- Date/Time : The date and time of the trip
- Lat : latitude of the pickup
- Lon : longitude of the pickup
- Base : code affiliated with the pickup [8]

There are some of questions this dataset can be used to answer, we'll choose the following:

- Uber trips and distribution
- Time when Uber trips occurs regularly
- Days when uber trips show up regularly
- Uber trips distribution in the Zones

3.1. Essential Packages

Some of the necessary libraries of R that we will use are –

- *ggplot2* - It is a plotting bundle that makes it easy to create complicated plots from records in a records frame. It offers a extra programmatic interface for specifying what variables to plot, how they are displayed, and well-known visible properties. Ggthemes
- *lubridate* - It is an R bundle that makes it less difficult to work with date and time.
- *dplyr* - A fast, constant device for working with information body like objects, each in reminiscence and out of memory.
- *tidyr* - It is new package deal that makes it handy to "tidy" your data. Tidy facts is statistics it truly is effortless to work (with *dplyr*)
- *DT* - Data objects in R can be rendered as HTML tables the usage of the JavaScript library 'DataTables' (typically with the aid of *R Markdown* or *Shiny*).
- *Scales* - Graphical scales map records to aesthetics, and supply strategies for robotically finding out the labels for axes.

3.2. Plotting the uber pickup trips by the hours in a day

In this section, we will use the ggplot feature to plot the wide variety of journeys that the passengers had made in a day. We will additionally use dplyr to mixture our data. In the ensuing visualizations, we can apprehend how the quantity of passengers fares for the duration of the day. We take a look at that the variety of journeys are greater in the nighttime round 5:00 and 6:00 PM. (see table 1)

Table 1. Datatables for uber pickup trips by the hours in a day.

Hour	Description
0	10386
1	67227
2	45865
3	48287
4	55230
5	83939
6	143213
7	193094
8	190504
9	159967
10	159148
11	165703
12	170452
13	195877
14	230625
15	275466
16	313400
17	336190
18	324679
19	294513
20	284604
21	281460
22	241858
23	169190

3.3. Distribution of pickups by The Hours of the Day

Now it's time to locate out what time most journeys have been made. Let's begin through discovering a everyday view of what time most journeys had been made. (see figure 2) From a general view of the data, looks like journeys peak between 6–8am and then peak again even more from 3pm and rises steadily until its on its peak at around 5pm before the pickups steadily drop till late night. (see figure 2)

3.4. Plotting uber trips by hour and month

From the above plot, we can easily see how there was once greater pickups in the month of September. We examine form the resulting visualization that the time increases, total number of pickups increases rapidly. We take a look at that the variety of journeys are greater in the nighttime round 5:00 and 6:00 PM in every month. (see figure 3)

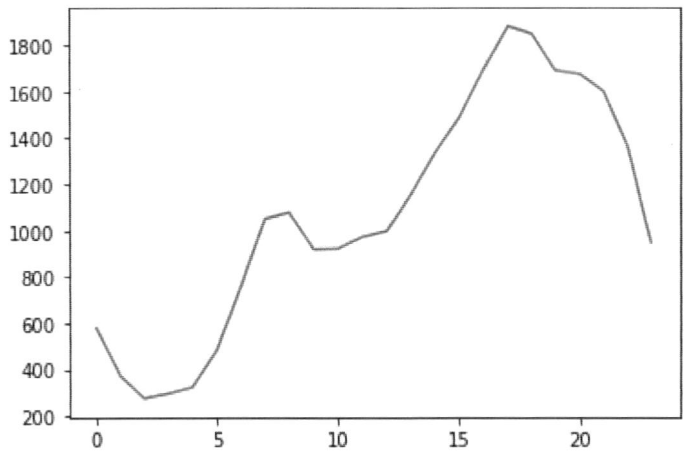

Figure 2. uber pickup trips by the hours in a day

3.5. Plotting uber data by trips during every day of the month
In this section, we will analyze how to plot our facts based totally on each and every day of the month. We examine from the resulting visualization that 30th of the month had the perfect journeys in the year which is more often than not contributed with the aid of the month of April. (see figure 4)

3.6. Number of uber Trips taking place during months in a year
In this section, we will visualize the range of journeys that are taking place every month of the year. In the output visualization, we look at that most journeys have been made for the duration of the month of September. Furthermore, we additionally gain visible reviews of the range of journeys that had been made on each day of the week. (see table 2)

Table 2. Number of uber Trips taking place during months in a year.

Month	Total
Apr	564516
May	652435
Jun	663844
Jul	796121
Aug	829275
Sep	1028136

From the above data table, we can easily see how there was once greater pickups in the months of july, august and September and highest pickups in the month of September.

3.7. Finding out the number of uber Trips by month and day
In the following visualization, we plot the range of journeys that have been taken through the passengers from every day of the month. (see figure 5)

IOP Conf. Series: Materials Science and Engineering 1119 (2021) 012013 doi:10.1088/1757-899X/1119/1/012013

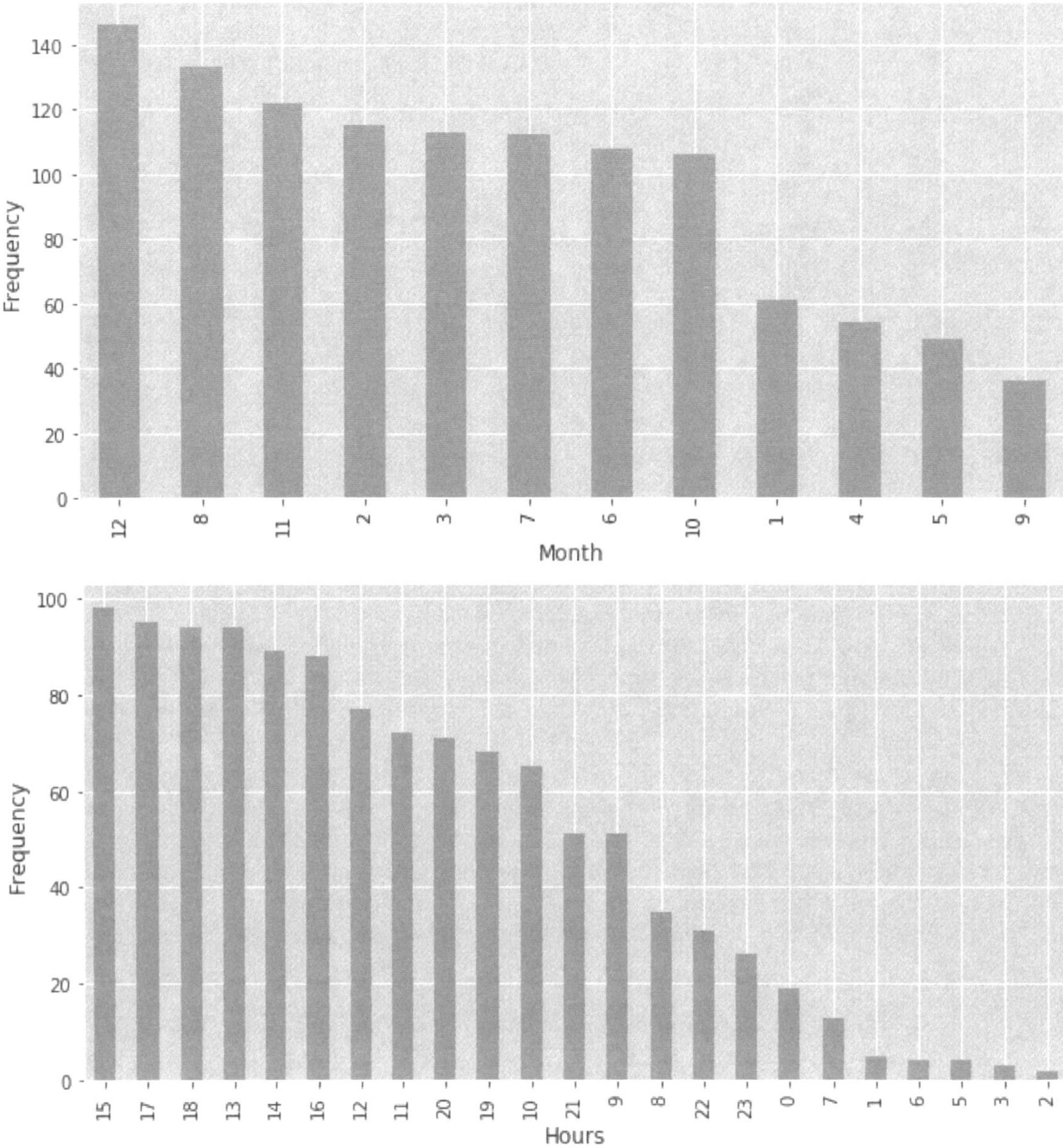

Figure 3. uber trips by hour and month

3.8. Finding out the number of uber Trips by bases

In the following visualization, we plot the range of journeys that have been taken through the passengers from every of the bases. There are 5 bases in all out of which, we study that B02617 (Weiter) had the easiest range of trips. Furthermore, this base had the best variety of journeys in the month B02617. Thursday found best journeys in the three bases – B02598 (Hinter), B02617 (Weiter), B02682 (Schmecken). From our data, there are 5 bases and every one of them has a unique region code that is used to become aware of it. To make it simpler for us to perceive the bases, we'll change the base codes on the Base column with the authentic names of the bases. Now let's create a plot to locate out the distribution of the pickups amongst the bases over these

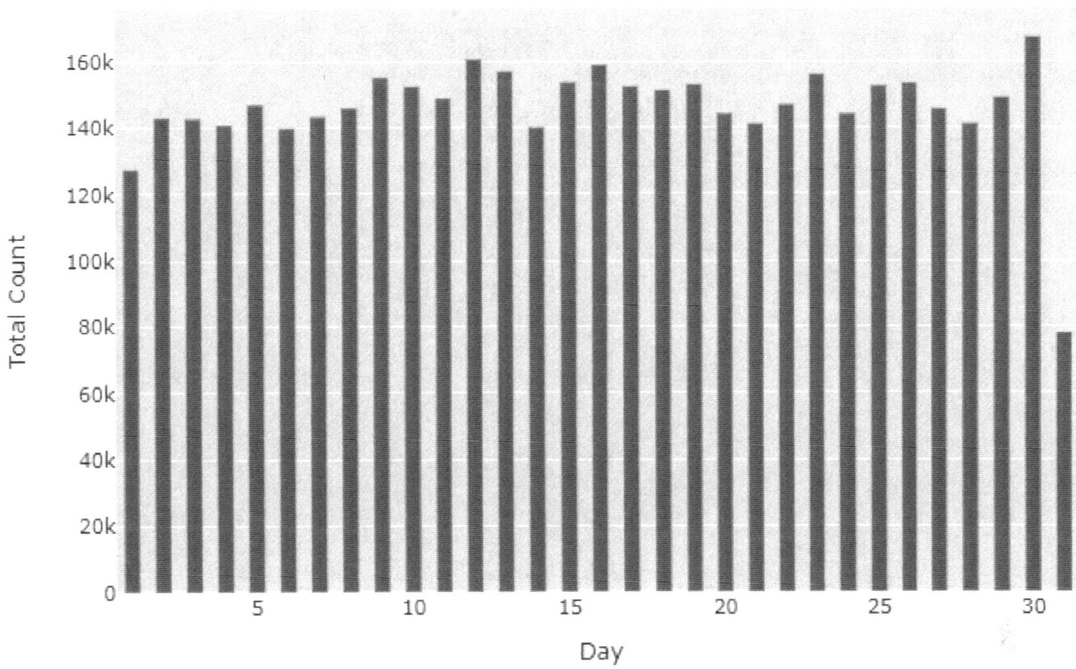

Figure 4. uber data by trips during every day of the month

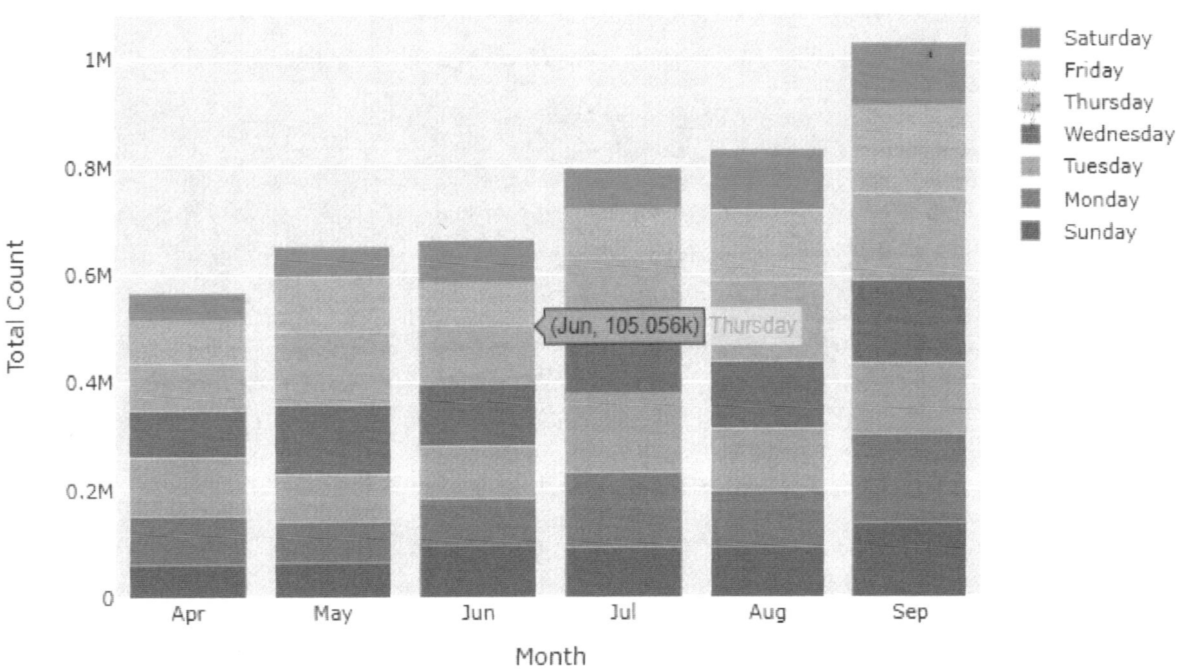

Figure 5. uber Trips by day and month

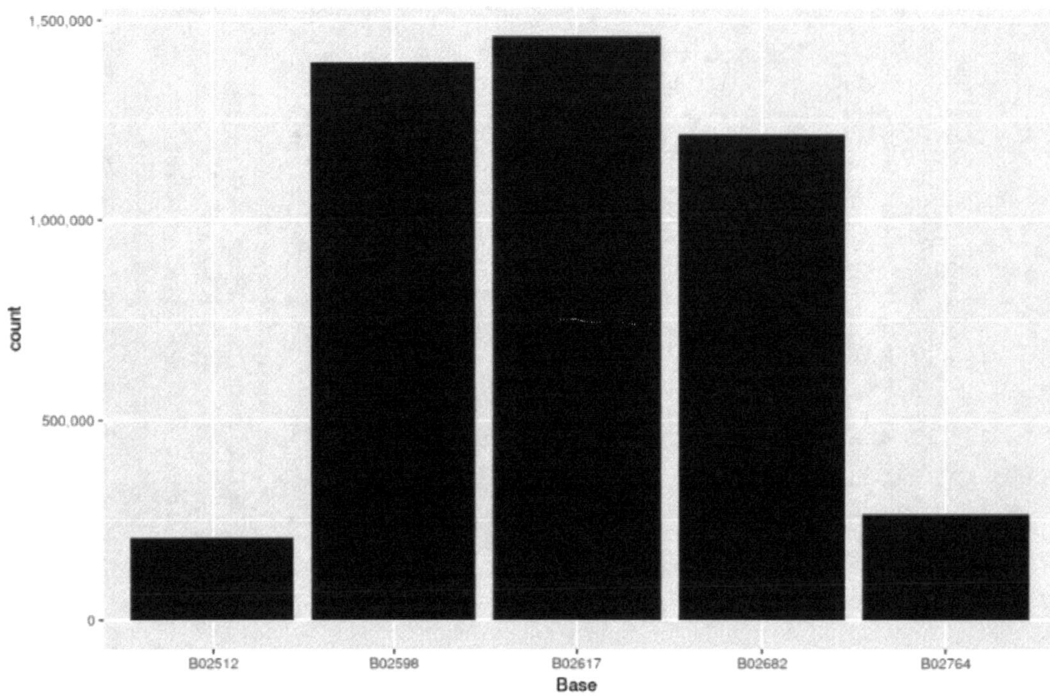

Figure 6. Number of uber Trips by bases

months [9]. From the above plot (see figure 6), we can easily see how there was once greater pickups. That's possibly due to the fact these areas are in quite populated areas. Under had the least quantity of pickups at less than 0.2 million.

3.9. Plotting uber data trips by bases and month
In the following visualization, we plot the journeys that have been taken by bases and month. From the plot 7, we can easily see how there was once greater pickups in Weiter (B02617) in the months of July, August and September. Bases Danach (B02764) was lowest pickups in comparison to all other bases but in the month of august, pickups increase rapidly.

4. Conclusion
At the end of the Uber data evaluation, the dataset we used for this undertaking protected statistics of Uber cars' ridership in the metropolis of New York for the six months of 2014. As we used to be exploring it, we seen that, in opposition to my preliminary intuition, the climate variables had now not any or very vulnerable effect on the ridership. Going similarly in my evaluation it used to be getting extra clear that the demand follows unique patterns each in the course of the day and at some stage in the week. We found how to create records visualizations. We made use of programs like ggplot2 that allowed us to plot a number of sorts of visualizations that pertained to a number of time-frames of the year. With this, we should conclude how time affected client trips. Finally, we made a geo plot of New York that furnished us with the small print of how a variety of customers made journeys from one of a kind bases.

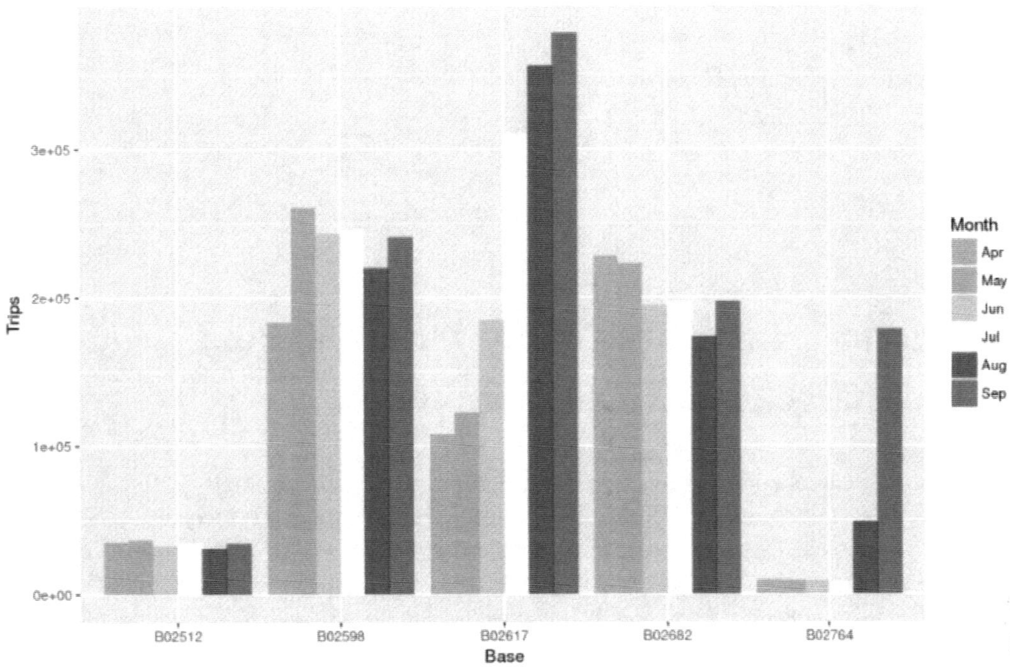

Figure 7. uber data trips by bases and month

References

[1] Wang C, Hao P, Wu G, Qi X and Barth M 2018 Predicting the number of uber pickups by deep learning Tech. rep.
[2] Yaniv A 2020 Big Data; Uber Forecast In New York | by Aviv Yaniv | Towards Data Science https://towardsdatascience.com/big-data-uber-forecast-in-new-york-5a25ffb9914f (Accessed on 01/30/2021)
[3] Faghih S, Safikhani A, Moghimi B and Kamga C 2019 *Journal of Computing in Civil Engineering* **33** 05019002
[4] Shah D, Kumaran A, Sen R and Kumaraguru P 2019 *Companion Proceedings of The 2019 World Wide Web Conference* pp 130–136
[5] Ahmed M, Johnson E B and Kim B C 2018 *Available at SSRN 3267082*
[6] Verma N and Baliyan N 2017 *2017 8th International Conference on Computing, Communication and Networking Technologies (ICCCNT)* (IEEE) pp 1–7
[7] FiveThirtyEight 2019 Uber Pickups in New York City | Kaggle https://www.kaggle.com/fivethirtyeight/uber-pickups-in-new-york-city (Accessed on 01/30/2021)
[8] TaG M 2018 Exploratory data analysis for Uber trips | Kaggle https://www.kaggle.com/mohamed08/exploratory-data-analysis-for-uber-trips (Accessed on 01/30/2021)
[9] Omondi E 2020 Exploratory Analysis Of Uber Pickups In NYC from April to September 2014 | by Elvis Omondi | Analytics Vidhya | Medium https://medium.com/analytics-vidhya/exploratory-data-analysis-of-uber-pickups-in-nyc-from-april-to-september-2014-ec78d5a35f78 (Accessed on 01/30/2021)

Optimization of thermal oxidation for 4H-SiC and fabricate/characterize MOS capacitor

Rajesh Singh Shekhawat[a], Nongthombam Joychandra Singh[b], Moh. Shakil[b], Prashant Kumar[b], Dhirendra Kumar[b], Prem Kumar[b], Arvind Kumar Singh[b], Prateek Kothari[b], Surajit Das[b], Priyavart Prajapat[b], Deepak Kumar Panwar[b], Sanjeev Kumar[b], T Eshwar[b], Sumitra Singh[b,c] and Dheerendra Singh[a]

[a]Department of EEE, Birla Institute of Engineering and Technology, Pilani-333031 India,

[b]Semiconductor Device Design Group, CEERI-CSIR Pilani-333031, India,

[c]Academy of Science and Innovative Research (AcSIR), CSIR-CEERI Campus, Pilani-333031, India

Email: p20170503@pilani.bits-pilani.ac.in, joychandra96@gmail.com, mohdshall5201@gmail.com, shprashant.2008@rediffmail.com, dhirendr@ceeri.res.in, prem@ceeri.res.in, arvind@ceeri.res.in, kothari@ceeri.res.in, surajit@ceeri.res.in, priyavart@ceeri.res.in, deepak_panwar145@ceeri.res.in, sanjeevkumar@ceeri.res.in, eshwar@ceeri.res.in, sumitra@ceeri.res.in, dhs@pilani.bits-pilani.ac.in.

Abstract. On the 4H-SiC substrate, C-face and Si-face oxide layers have been grown by thermal oxidation process and sputtering. The thermal oxidation temperature dependence of 4H-silicon carbide (SiC) is systematically investigated using capacitance-voltage (C–V) measurements. The oxidation quality and thickness vary according to the temperature and time duration of the thermal oxidation. The layers' thicknesses are determined by atomic force microscopy (AFM), and the temperature range is between 800°C and 1110 °C. The primary reason to fabricate the Metal-Oxide-Silicon (MOS) capacitor is to know the thermal oxidation process and a working principle. In this paper, we optimize a thermal oxidation process and fabricate the MOS structure. Then we determine the various parameters such as flat band voltage (V_{fb}), Inversion threshold voltage (V_t), Surface depletion capacitance (C_{dep}), Oxide capacitance (C_{ox}), the total capacitance of the device (C_o), doping concentration (N_d), Depletionwidth(X_d),Maximumdepletionwidth (X_{dt})andInterfacetrapdensity(D_{it}). Finally, we analyze and discuss the MOS capacitance.

Keywards- Fabrication, Thermal oxidation, Sputtering, Capacitance-voltage (C–V) measurements.

1. Introduction

In VLSI Technology, semiconductors can be oxidized by different methods like, e.g., plasma-enhanced chemical vapor deposition (PECVD), electrochemical anodization, and thermal oxidation. To grow top-quality oxide (SiO_2) layer on a Si wafer, various grown methods exist, and continuous research is going on. Thermal oxidation is a way the foremost importance for device fabrication. Same as the Si-technology, Silicon Carbide (SiC) is a single compound semiconductor that is thermally oxidized in the form of SiO_2. This suggests that we will directly try to the enlargement of SiC-Technology.

Several researchers and various groups are exceptionally performing on the growth mechanism of SiO_2 on the SiC substrate surface beside the Si growth mechanism. Since a compound semiconductor, SiC material having Si and C atoms, C atoms' character is detected to be very critical during the thermal growth of the SiO_2. Several experiments [7-9] verify that the MOS-structure interface and dielectric properties are directly affected by the availability of the C species during the thermal grown oxidation [10]. To understand and control the electrical characteristics of the SiC-based device, the nature of thermally grown SiO_2 on SiC wafer plays an important role; this is one reason various researchers rigorously work in this area. According to the reports, the comparison with the Si substrate growth rate of oxide on the SiC substrate is much lesser [11-13]. In equivalent oxidation condition speed of reaction on the Si-wafer is far faster than of SiC-wafer.

It has been also noted that a remarkable phenomenon occurred that the face termination oxidation on SiC substrate surfaces, which indicates that different oxidation thickness has on both the polar faces (Si and C face). This thickness, quality of oxidation layer, and oxidation rate change upon the polytypes and crystal orientation of SiC.

The metal-oxide-semiconductor (MOS) structure is the core of MOS technology. MOS structure (customarily known as a MOS capacitor) is connected as a two-terminal device, with one electrode connected to the metal and, therefore, the other electrode attached to the semiconductor. As a result, the voltage-dependent capacitance characteristic is drawn. During this paper, we first consider constructing a MOS capacitor then drawn the capacitance-voltage (C-V) relationships, which can utilize in the development of the MOS transistor model [26].

2. 4H-SiC substrate Specification

Table 1. 4H-SiC substrate Specification for fabrication

Parameter	Value
Substrate	
Substrate diameter	3 inches
Substrate material	SiC-4H single crystal wafer
Substrate wafer orientation	0001 with 4-degree off axis
The thickness of the substrate wafer	350 μm
Substrate face	Silicon Face
Substrate doping type	n-type
Resistivity of substrate	0.015-0.028 Ωcm
Polishing of substrate	Chemo Mechanically Polished, Double side polished
Micro-pipe density in substrate	$\leq 1/cm^2$
Epilayer 1 (buffer layer)	
Epilayer 1 doping type	n-type
Epilayer 1 doping density	10^{18}/cc
Epilayer 1 thickness	1 μm
Epi layer 2	
Epilayer 2 doping type	n-type
Epilayer 2 doping density	$8X10^{14}$/cc
Epilayer 2 thickness	50 μ m

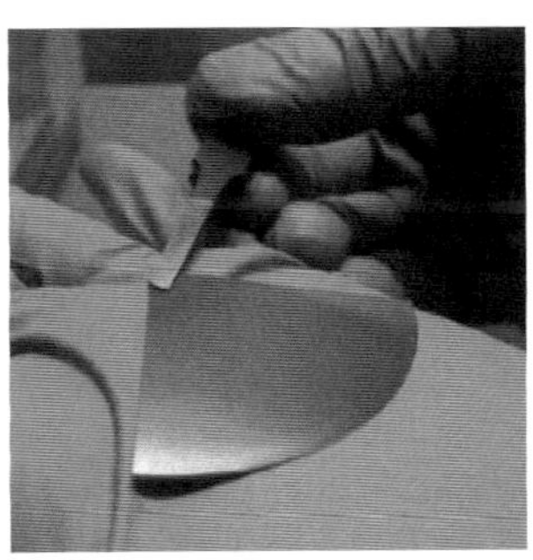

Figure 1. The schematic picture of quarter pieces of 4H-SiC substrate.

M/s CREE Research Inc USA provides the Single crystalline SiC-wafer. In our paper, an n-type 4H-SiC material was with small doping concentration and the epitaxy layer thickness (50 μm) on Si-face.

3. Thermal oxidation Furnace

The thermal oxidation furnace was used for the thermal growth of SiO2 on the 4H-SiC substrate, as shown in figure 3. It is usually contained of following parts: heating zone (consists of several heating coils), measurement system, cabinet, temperature controller, horizontal quartz tubes (the place where the wafers experience oxidation), loading station (used for loading or unloading) wafers into (or from) the process tubes and a gas flow meter (for observing used gases flow into and out from furnace). There are mainly three quartz tube heating zones, i.e., left-hand, right-

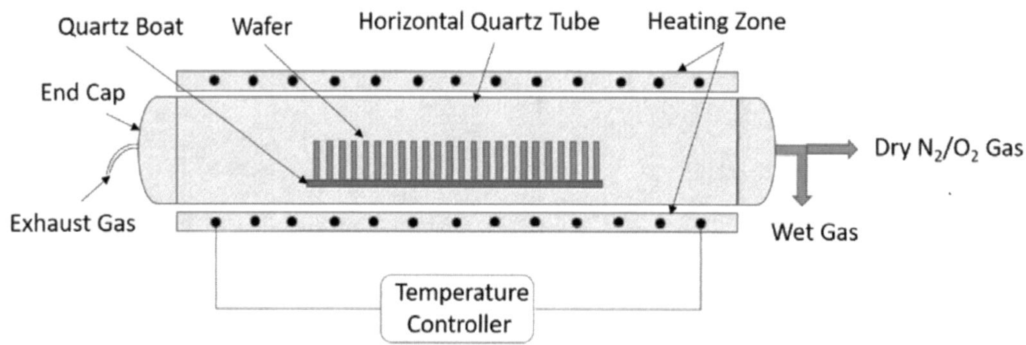

Figure 2. Schematic diagram of horizontal oxidation furnace.

hand, and middle. The temperature 4000±500 °C was fixed for the left-hand and right-hand zone at the complete process. There is three digital control system to increase and decrease the furnace temperature for all three different zones. The furnace is having two gas pipelines and one gas flow controller (MATHESON's). Too pure DI-water has been produced by quartz bubbler, and the temperature of the bubbler can be controlled by heating metal. In the quartz furnace, various gases (Wet O_2/dry O_2 or dry N_2) have been passed with the help of a quartz nozzle [27].

4. Cleaning process of the wafer

The SiC wafer clean-up technique is the same as Silicon, and all substances utilized in the wet-chemical process remained MOS grade. The SiC-wafer was a three-step chemical cleaning method like the Si-wafer, i.e., Degreasing, the standard Radio Corporation of America (RCA), and Piranha. Degreasing has three clean-up stages. In the first stage, to remove grease particles from the wafer's surface, wafers were dropped and boiled for 10 minutes in Trichloroethane (TCE). In the second stage, to remove light meatal ions or particles, the wafers were dropped and boiled for 10 minutes in acetone. In the third stage, the wafers were dropped and boiled for 10 minutes in methanol. Finally, with the help of de-ionized (DI) water, the wafers were cleaned.

The standard Radio Corporation of America (RCA) clean-up method has two phases, the first phase is known as standard cleaning-I (SC-I), and the second stage is known as standard cleaning-I (SC-II). In the SC-I at room temperature, the wafers were dipped for 10 min in a high pH alkaline mixture (NH_4OH/H_2O_2/DI-water) within the relative amount of (1:1:5). SC-I is employed to remove the organic substances on the wafer, exposing the surface and remove hydrous oxide film. Then the wafer was clean in DI-water and dipped for one minute in hydrofluoric (HF) acid. The SC-II, at room temperature, the wafer was dipped in a combination of (HCl/H_2O_2/DI-water) within the relative amount of (1:1:6). Finally, clean in DI-water and with the 10 percent HF solution remove the native oxide. The SC-II cleaning is used to remove water-insoluble hydro oxide compounds, any dual-trace metals, and alkali ions disrobed by SC-I. The last cleaning treatment is Piranha cleaning. The wafers are

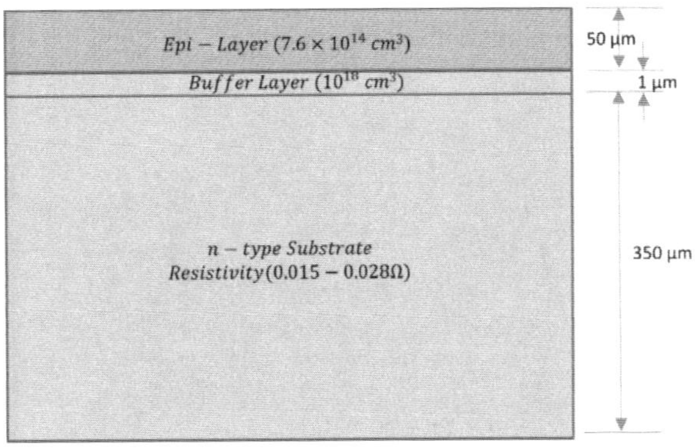

Figure 3. Schematic details 3-inches diameter 4H-SiC Substrate.

dipped for 15 minutes in piranha solution (H_2SO_4 and H_2O_2 within the ratio of 4:1) to remove any heavy metal resident on the wafer surface. After All, using DI-water, the wafers were clean-up and dipped in the 10 percent HF. The temperature ranging from 800 °C to 1110 °C was used for thermal oxidation procedure for various oxidation time. It tried to find the precise performance of thermal oxidation on both sides of SiC-wafer in both dry and wet oxidizing environments.

5. Oxidation Process

The thermal oxidation process temperature ranges from 800 °C to 1110 °C for different periods for both the process, as shown in figure 4 and figure 5. The wafers were placed in the middle zone of the furnace in the boat (quartz glassware), which carries multiple wafers. The oxidant process was started, and diffusion takes place. In the wet oxidation process, the quartz bubbler's temperature is always kept at a constant 850°C. The helical path has controlled the wet molecular oxygen (0.4 LPM) during the process. The wafers were loaded and

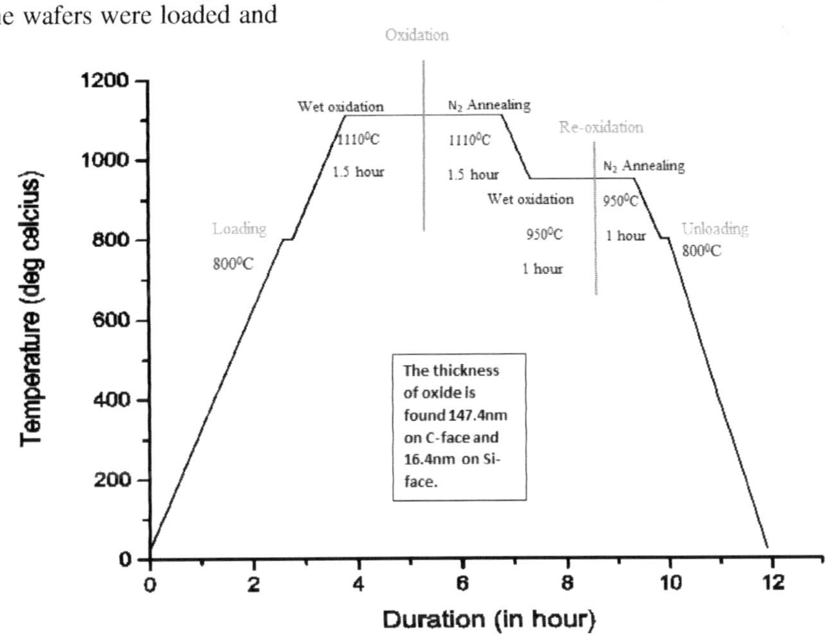

Figure 4. Process flow of wet thermal oxidation for 1.5 hour.

unloaded in the 1.9 LPM flow of nitrogen at 800 °C, increasing and decreasing the furnace's temperature by 5 °C/min, which is shown in Figures 4 and 5. In the first process (figure 4), oxidation (wet oxidation

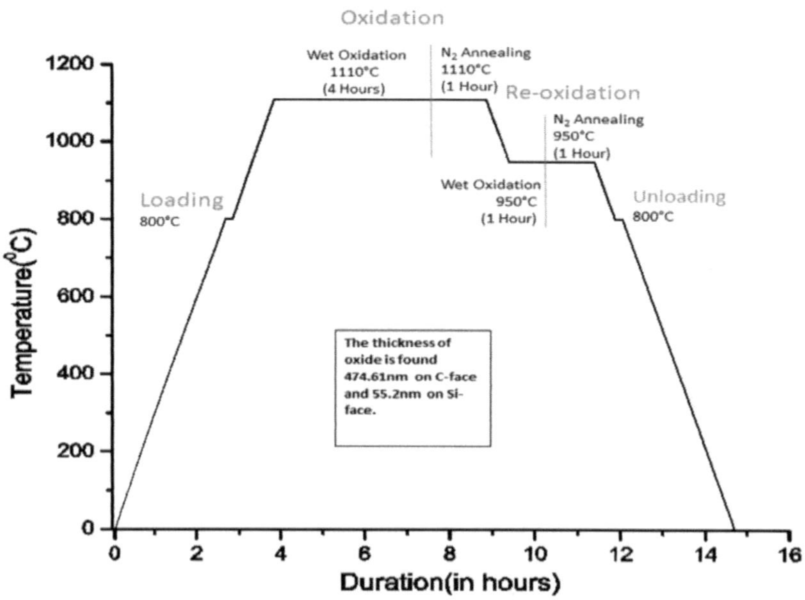

Figure 5. Process flow of wet thermal oxidation 4 hours.

and annealing duration is 1.5 hr each) for 3hrs at 1110 °C and re-oxidation (wet oxidation and annealing duration is 1 hr each) for 2 hr at 950 S. In the second process, as shown in figure 5, oxidation (wet oxidation and annealing duration is 4 hrs and 1 hr respectively) for 5 hrs at 1110 °C and re-oxidation (wet oxidation and annealing duration is 1 hr each) for 2 hr at 950 °C. The thickness of oxide was measured by the ellipsometry technique with verification of DAKTEK surface profiler on both end faces. The thermally grown oxide layer thickness is found 147.4nm on C-face and 16.4 nm on Si- face (Process flow of wet thermal oxidation for 1.5 hours). The thickness of thermally grown oxide layer oxide is 474.61 nm on C-face and 55.2 nm on Si-face (Process flow of wet thermal oxidation 4 hours). Wet thermal oxidation of 4H-SiC is depended on various conditions such as temperature, pressure, and time.

6. Fabrication of MOS Capacitor

The structure of a MOS capacitor is demonstrated in Figure 6. MOS capacitor consists of the oxide-

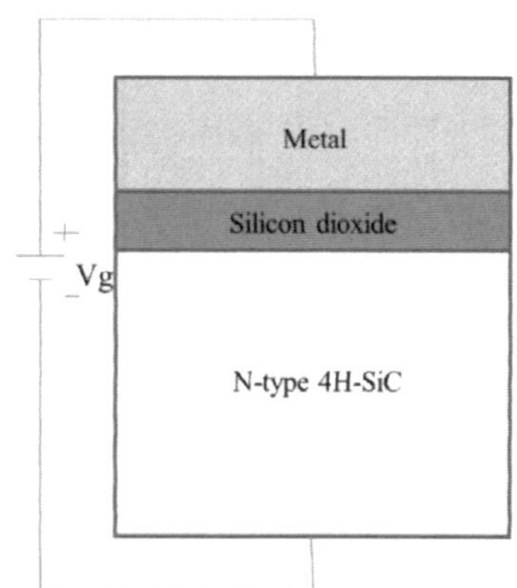

Figure 6. Structure of MOS capacitor.

coated n-type 4H-SiC wafer. The oxide layer (silicon dioxide SiO_2) is grown by thermal oxidation, and after the oxide layer, a conducting layer (Ni) (called a gate) is grown by sputtering. Through a metal defining mask, the gate structure is defined the metal is vaporized on the wafer. The ohmic contact is obtained by evaporating the metal on back (bulk) contact, which is in intimate contact with SiC. Vacuum annealing of 1hrs at 300°C is done before and after contact deposition.

Table 2. Specification of parameter

Vacuum-4x10^{-7} (during deposition)	Thickness-2000Å
Thickness-300 Å for Ti	Deposition Rate-1.4 Å/sec
-2000 Å for Au	Rev Time-4.5
Deposition rate- 0.4 Å/sec for Ti	Power -100 W(DC)
-1.7 Å /sec for Au	Frequency-100kHz
Time-24 min for Ti	Argon-7.0 sccm
-60 min for Au	RF-13.56 MHz

The MOS capacitor structure useful for studying the silicon surface, and the appropriate MOS transistors and C-V characteristics are employed to examine IC fabrication method. For this justification, the MOS capacitor as a test device is used to measure HF and LF C-V plots, which measure gate oxide thickness t_{ox} oxide charges and interface-state density D_{it}, flat band voltage V_{fb}, doping concentration etc. [1]-[3].

7. Analysis of C-V characteristics

Across the device, by applying the dc bias from -40 V to 1 V at 1 MHz frequency, the C-V characteristics were measured using a 4284A LCR meter. Through the LabVIEW software, data has been carried out through computer-aided measurement facilities. As shown in figure 8, the characteristics of the fabricated structure of applied voltage with capacitance at a different frequency, and figure 9 show the different voltage and $1/C^2$. Various parameters have been calculated by applying appropriate equations with the help of the drawn C-V characteristics. Characteries the MOS capacitor and find the various parameter like flat band voltage V_{fb}

Figure 7. Fabricated Final Device

IOP Conf. Series: Materials Science and Engineering 1119 (2021) 012014 doi:10.1088/1757-899X/1119/1/012014

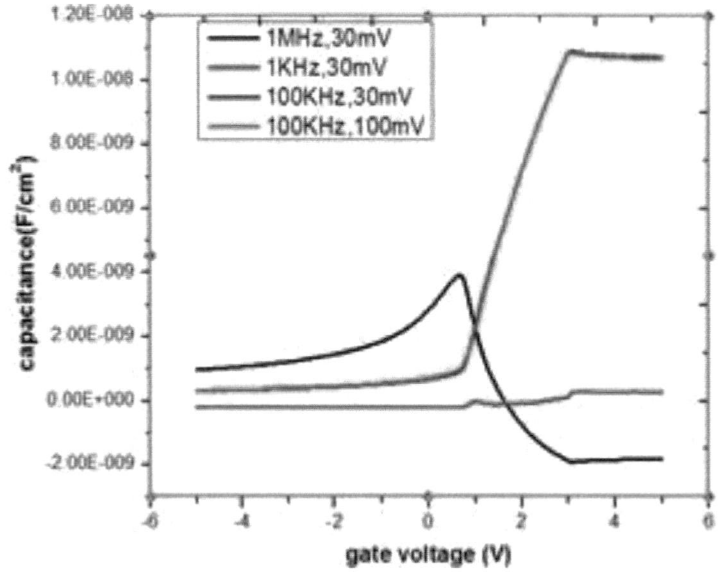

Figure 8. Measured characteristic curves of applied bias voltage vs. capacitance at a different frequency.

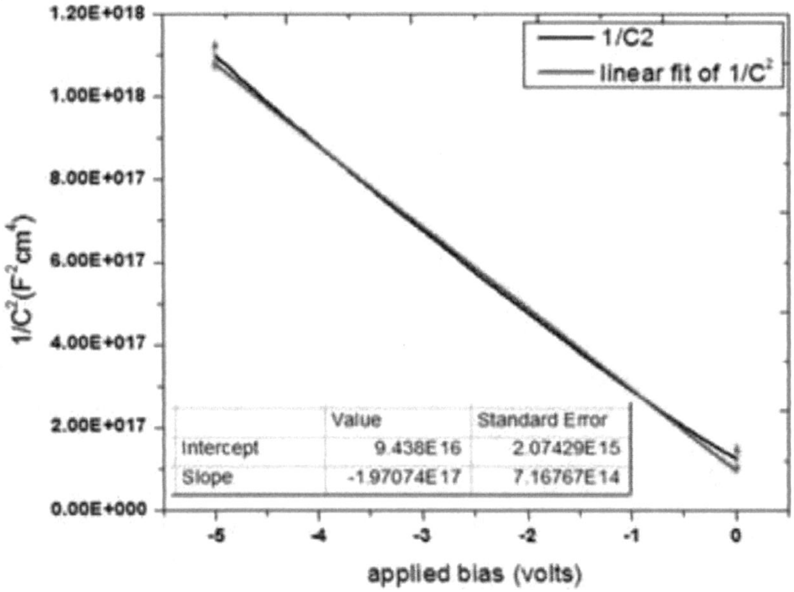

Figure 9. Applied Voltage vs $1/C^2$.

1.09V, Inversion threshold voltage V_t 2.66 V, Surface depletion capacitance C_{dep} 68.67 nF/cm^2, Oxide capacitance C_{ox} 115 nF/cm^2, Total Capacitance of the device C 0.7482 X 10^{-18}F, doping concentration N_d 7.69 x10^{14} CC, Depletion width X_d 1.37 x10^{-4} cm, Maximum depletion width X_{dt} 1.937x10^{-4}cm and Interface trap density D_{it} 3.6x10^{-9} states/cm^2 are calculated.

8. Conclusions

This paper represents an optimized 4H-SiC thermally grown oxide mechanism through an experiment. Based on experimental results achieved, various conclusions have been discovered, such as A face terminated oxidation performance has been studied, which suggests that the oxidation growth rate on Si-face is slower than C-face. In the wet oxidation, Si-face oxide thickness is 8 to 12 times slower than the C-face. Finally, we successfully fabricated and characteries the MOS capacitor and found the device's various related parameters.

Acknowledgments
The authors would like to thank all the Semiconductor Device Design Group staff members for their enormous support during the batch fabrication of diode structures. The authors are thankful to Director CSIR-CEERI for his continuous support and encouragement. Financial support from DRDO through project no CLP 6555 is acknowledgeable.

References

[1] Vickridge I, Ganem J, Hoshino Y and Trimaille I 2007 Growth of SiO_2 on SiC by dry thermal oxidation: mechanisms *J. Phys. D: Appl. Phys.* **40** 6254-63.

[2] Hiroshi Yano, Fumito Katafuchi, Tsunenobu Kimoto, and Hiroyuki Matsunami 1999 Effects of Wet Oxidation/Anneal on Interface Properties of Thermally Oxidized SiO2/SiC MOS System and MOSFET's *IEEE Trans. Electron Devices* **46** No. 3 504-10.

[3] Kamimura K, Kobayashi D, Okada S, Mizuguchi T, Ryu E, Hayashibe R, Nagaume F and Onuma Y 2001 Preparation and characterization of SiO_2/6H-SiC metal insulator semiconductor structure using TEOS as source materials *Appl. Surf. Sci.* **184** Issue 1-4 346-49.

[4] Lai P T, Xu J P, Wu H P, Chen C L 2004 Interface properties and reliability of SiO_2 grown on 6H-SiC in dry O_2 plus trichloroethylene *Microelectron. Reliab.* **44** Issue 4 577-80.

[5] Meakawa M, Kawasuso A, Chen Z Q, Yoshikawa M, Suzuki R, Ohdaria T 2005 Structural defect in SiO_2/SiC
interface probed by a slow positron beam *Appl. Surf. Sci.* **244** Issue 1-4 322-25.

[6] Zetterling C, Ostling M, Harris C I, Wood P C, Wong S S 1998 UV-ozone precleaning and forming gas annealing applied to wet thermal oxidation of p-type silicon carbide *Mater. Sci. Semicond. Process.* **2** Issue 1 23-27.

[7] Eckhard Pippel, Jorg Woltersdorf, Halldor O Oafsson and Einar O Sveinbjornsson 2005 Interfaces between 4H-SiC and SiO_2: Microstructure, nanochemistry, and nearinterface traps *J. Appl. Phys.* **97** 034302.

[8] Mark Schürmann, Stefan Dreiner, Ulf Berges, and Carsten Westphal 2006 Investigation of carbon contaminations in SiO2 films on 4H-SiC (0001) *J. Appl. Phys.* **100** 113510.

[9] Chen X D, Dhar S, Isaacs-Smith T, Williams J R, Feldman L C, and Mooney P M 2008 Electron capture
and emission properties of interface states in thermally oxidizedm and NO-annealed SiO_2 /4H-SiC *J. Appl. Phys.* **103** 033701.

[10] Vathulya V R, Wang D N and White M H 1998 On the correlation between the carbon content and the electrical quality of thermally grown oxide on p-type 6H-Silicon Carbide *Appl. Phys. Lett.* **73**, 2161-63.

[11] Vickridge I C, Ganem J J, Battisig G, and Szilagyi E 2000 Oxygen isotropic tracing study of the dry thermal oxidation of 6H-SiC *Nucl. Instrum. Methods Phys. Res.* **161**B 462- 66.

[12] Song Y, Dhar S, Feldman L C, Chung G, and Williams J R 2004 Modified deal and grove model for the thermal oxidation of silicon carbide *J. Appl. Phys.* **95** 4953-57.

[13] Knaup J M, Deak P, Frauenheim T, Gali A, Hajnal Z, and Choyke WJ 2005 Theoretical study of the mechanism of dry oxidation of 4H-SiC *Physical Review* B**71** 235321-28.

[14] Schuermann M, Dreiner S, Berges U and Westphal C 2006 Structure of the interface between ultra thin SiO_2 film and 4H-SiC (0001) *Physical Review* B**74** 035309-13.

[15] Fiorenza P and Raineri V 2006 Reliability of thermally oxidized SiO_2/4H-SiC by conductive

atomic force microscopy *J. Appl. Phys.* **88** 212112-15.

[16] Deal B E and Grove A S 1965 General relationship of the thermal oxidation of silicon *J.Appl. Phys.* **36**, 3770.

[17] Eckhard Pippel and Woltersfordf 2005 Interface between 4H-SiC and SiO_2; microstructure, nanochemistry and interface traps *J. Appl. Phys.* **97** 034302.

[18] Schmeiber D, Batchelor D R, Mikolo R P, Halfmann O and L-Spez A 2001 Oxide growth on SiC (0001) surfaces *Appl. Surf. Sci.* **184** Issue 1-4 340-45.

[19] Htun Aung M T, Szmidt J and Bakowski M 2002 The study of thermal oxidation on SiC surface", *J Wide Bandgap Material* **9** No. 4 313-18.

[20] Kosugi R, Fukuda K and Arai K 2003 Thermal oxidation of (0001) 4H-SiC at high temperature in ozone- admixed oxygen gas ambient *Appl. Phys. Lett.* **83** 884.

[21] Nicollian E H and Brews J R 1981 *MOS (Metal Oxide Semiconductor) Physics and Technology* (John Wiley and sons, New York) p. 673.

[22] Barbottin G B and Vapaille A 1989 Eds, *Instabilities in Silicon Devices* (North-Holland, New York) Vols I and II.

[23] Schroder D K 1990 *Semiconductor Material and Device Characterization* (John Wiley and Sons, New York).

[24] S. M. Sze 1982 *Semiconductor Physics and Technology* (John Wiley and Sons, New York).

[25] Arora N 1993 *MOS Capacitor. In: MOSFET Models for VLSI Circuit Simulation. Computational Microelectronics* (Springer, Vienna) https://doi.org/10.1007/978-3-7091-9247-44.

[26] Sanjeev Kumar Gupta and Jamil Akhtar 2011 *Thermal Oxidation of Silicon Carbide (SiC) Experimentally Observed Facts, Silicon Carbide* (IntechOpen, Rijeka) chapter-9 doi- 10.5772/20465, https://doi.org/10.5772/20465.

[27] Fiorenza P, Giannazzo F, Roccaforte F 2019 Characterization of SiO_2/4H-SiC Interfaces in 4H-SiC MOSFETs: A Review *Energies* **12** 2310.

A Novel and Efficient Approach for Content- Based Image Retrieval

Akash Saxena[1], Vikram Rajpoot[2*], Ankur Chaturvedi[3]
[1]Professor & Head CITM College Jaipur Rajasthan, India
[2,3]Assistant Professor Department of CEA GLA University Mathura, India
[2*]vikram.rajpoot@gla.ac.in

Abstract: In content-based image retrieval (CBIR) means the quest of an image continues at the real content material of picture to a positive volume to its metadata. The CBIR System has utilized for the remove the features, indexing individual's features by inappropriate formation and capably given a response to user's uncertainty. To provide the significant response to client question, CBIR gave some run of work. A novel scheme is proposed in this research work for color picture indexing by introducing the ease of the SVM system. A novel method is introduced for color picture indexing by exploit for it easy of the SVM system. This feature Extraction was seen as the binary classification problem and SVM was used for the problem's solution. It is concluded that to attain the extraction at the high speed as well as for making it so flexible that it can also adjust with the images of large size. The objective of this paper is to stand for the importance of maintaining vector machine in the well-organized retrieval of an image in SVM is utilizes as classifier that performing task of classification the image and this process of classification are providing to all picture that are mined later by feature extraction process
Keywords: CBIR, Error diffusion, Vector Quantization, SVM, etc.

1. Introduction

Advances in the generation along with scanners, digital cameras, garage media and big online records have led to a propagation of each specialized and individual series of digital images (DI). Image retrieval (IR) failed in an in particular energetic observe vicinity during the last 10 years, however number one evaluation article on access approach in photo databases appear as already inside the near the beginning eighty decades [1]. The fundamental CBIR system includes important component, function drawing out and categorization. The photos are reserved in a record known as picture Database. Behind pre-processing, pictures are partitioned by means of utilizing for the approach clarification. CBIR device consists of important steps. The first one is the function withdrawal, in which a hard and fast of the characteristic is created to symbolize content of every picture [2].

The paintings targeted on the use of small level features like color, texture, and shape and spatial give an explanation for picture illustration. The most basic utilize of the time period CBIR in the text appear to have been utilized by an author in to explain their test of computerized recovery of photos as of a record by way of coloration and shape capabilities [4]. There are 4 levels of getting again the picture from the database primarily founded on the numerical assets of the enter inquiry descriptions border [5].

- picture database creation stage
- enter pictures from the scanner, thinning editing's Outline based pictures recovery stage
- Classify the worldwide outline matches Global hash table creation stage [6].

1.1 *Error Diffusion*

Its creation Floyd and Steinberg, the error diffusion (E-D) set of rules have attracted more consideration. The principal gain of the preliminary model of the set of rules is its simplicity and its right normal visible first-rate of the produced binary photograph. For these kinds of motives, it has become very famous in numerous pictures packages [7]. Unlike the block trade and ordered irresolute strategies, which deal with each pixel in my view, errors dissemination quantity every pixel used in the neighborhoods manner. In this example, the price of each output factor relies upon now not handiest at the fee of the matching enter point.

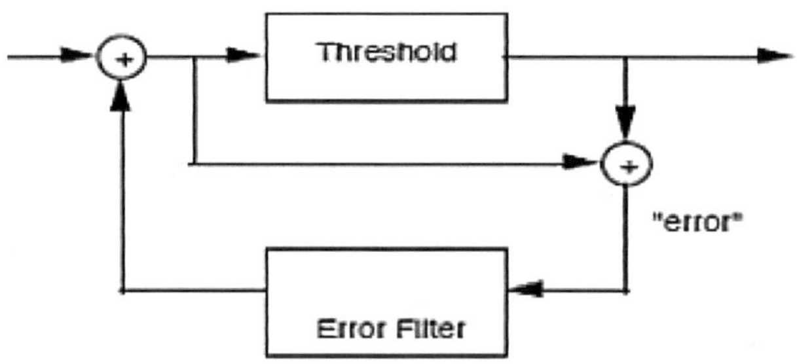

Figure.1 Error diffusion half toning method

1.2 *Vector Quantization*

- Vector Quantization is a simple method, normally however now not solely used, for data compression. It is applied for photograph firmness for a variety of years. VQ encompass separating a picture into quite a few equal sized vectors (or blocks), and every vector is the map with a codeword from the codebook to predetermine a photo. A vector Quantizer Q of proportions of K and N exact as arranging of K-d Euclidean area RK right into a constrained subgroup of Y of RK i.e.
- Q: RK →Y
- Where Y =(x_i ; i =1, 2, ,N) and x_i is the ith vector in Y
- Y is about called the VQ Codebook or VQ desk. This is a fixed of duplicate vectors of N, which means it has N wonderful basics every a vector in R K. At the encoder every vector x is in the region of the match with a password in a codebook, an index of the codeword is transmitted. Euclidean distance (ED) or a Manhattan detachment may be used for computing the first rate in shape codeword. Thus an encoded image containing the indexes is produced. At the decoder side, the indexes are map lower back to codebook and unique image is generated. The transmitter and receiver should have the same codebook. [8] To summarize there are three steps worried in Vector Quantization
- Generating Codebook: For the codebook generation, length of the vector wants to be decided. And divide the photograph into vectors. After this apply codebook producing set of rules.
- Encoding Image: Encoding means find the maximum exquisite healthy open sesame for

every vector and update the codeword with the directory of the matched codeword.

- Decoding Image: Decoding method using the similar codebook, for each index, update it with the respective codeword. And deliver collectively all codeword The parent beneath shows the running of VQ procedure

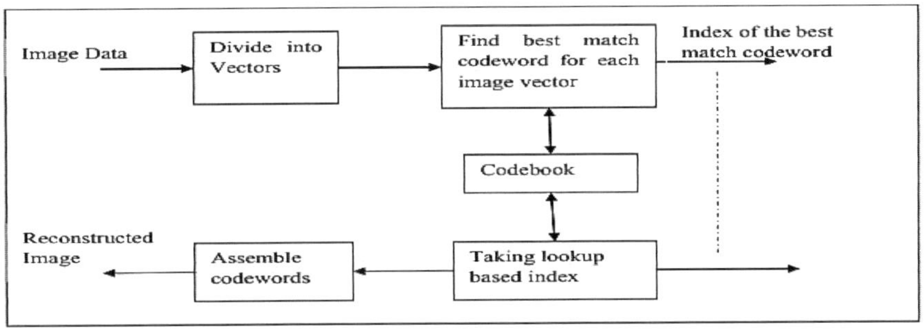

Figure.2 VQ encoding and decoding process

2. Literature review

Behzad Merhrbakhsh Choobar et al. [2017] suggest novel approach called local full-directional pattern in CBIR. Along with, instead of utilizing the method into picture itself, author utilize it to a novel picture obtained by finding the mean of 3™3 sub-partition gray value at every pixel's value. In the new approach, 8 different directions are given. Every pixel allot one of the directions related to its association with its adjacent neighbors, which is obtained utilizing first-order derivatives in vertical, horizontal and two diagonal directions. It utilized Corel 1000 dataset to evaluate this technique with LBP and neighborhood tetra sample (LTrP). This suggested approach represents better growth in average precision and recall. By utilizing the approach in the mean value of 9 pixel-windows, suggested approach represents significant retrieval results for noisy pictures [9].
Abdolreza Rashno et al. [2017]. In this paper, new CBIR) technique is suggested in neutrosophic (NS) area. In that, RGB pictures are first convert to 3 subsets in NS area and then partitioned. In every partition of a picture, color properties containing dominant color descriptor (DCD), histogram and statistic parts evaluated. All evaluated properties from any partitioned picture or the complete pictures are collected to make a feature vector. Feature vectors are supplied in ant colony optimization (ACO) function option which takes the most similar properties. Selected properties are utilized for last selected process. Suggested CBIR technique is calculated on Corel picture database. Implement consequences represent suggested approach better than previous approach (same characteristic vector and properties collection technique) by 2% and 1% with deference to precision and recall, in that order. Also, the suggested approach gets the increase of 13% and 2% in precision and recall, as compared to earlier approaches [10]. Devi S, et al. [2016] In this paper color pictures are the FE the usage listed on EDBTC. A novel structure of CBIR with unsupervised clustering is utilizing properly here wherein the amount of time required for comparing the goal and question photograph is drastically decreased. New end outcome indicates that the proposed technique now not most effectively accomplish an extraordinary of photograph reduction but get a large bargain in quantity of time compulsory for picture retrieval. [11]. Navneet Kaur et al.[2016] In this paper, importance opinion is used to the transaction with this trouble which founded on SVM gadget they expansively used inside the CBIR scheme to bridge the semantic space linking low downstage facial appearance and immoderate stage person interest capabilities. The knowledge procedures are predominantly applied for the categorization of the picture in labeled and unlabeled datasets. In our planned paintings

we need to artwork on KNN, SVM and Bays Classifier to categorize the pictures. [12]. Jing-Ming Guo et al. [2014] This paper gives a singular come near to index color picture utilizing the FE from the EDBTC. The EDBTC creates coloration quantizes and a chunk map picture that is further method makes use of Vector Quantization for producing a picture properties descriptor. Herein, functions are suggested, mainly Color Histogram property and Bit Pattern Histogram characteristics, to compute assessment among a question in picture and the aimed picture in the database [13].

3. Propose work

A novel procedure is given in this research for color image indexing thru exploiting simplicity of the SVM approach. This feature Extraction (FE) seen as the binary classification problem and SVM used for the solution of this problem. It is concluded that to attain the extraction at the high speed as well as for making it so flexible that it can also adjust with the images of large size.

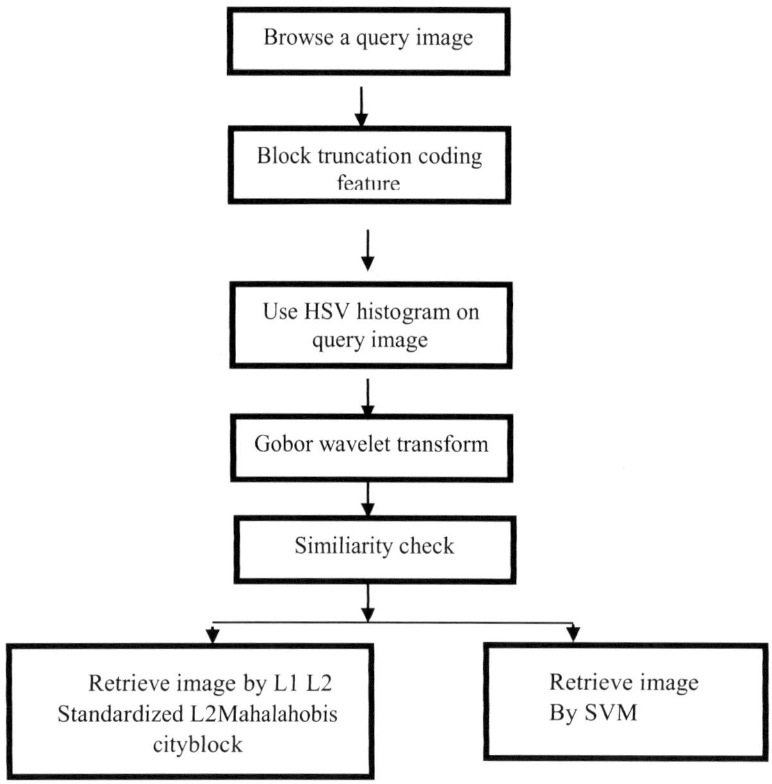

Figure.3 Flow Chart of Proposed Work

Figure.4 using image retrieval technique like L1.

The objective of this paper is to show the importance of support vector device in the effective collection of the photograph in this SVM is used because the classifier that is performing the task of classification image and process of classification is taken to the entire picture which is evaluated after FE process.

4. Results Analysis

Figure.5 Images retrieve using SVM

Table 1. Propose Table Precision and Recall

Parameter	Value
Precision	0.8400
Recall	0.7200

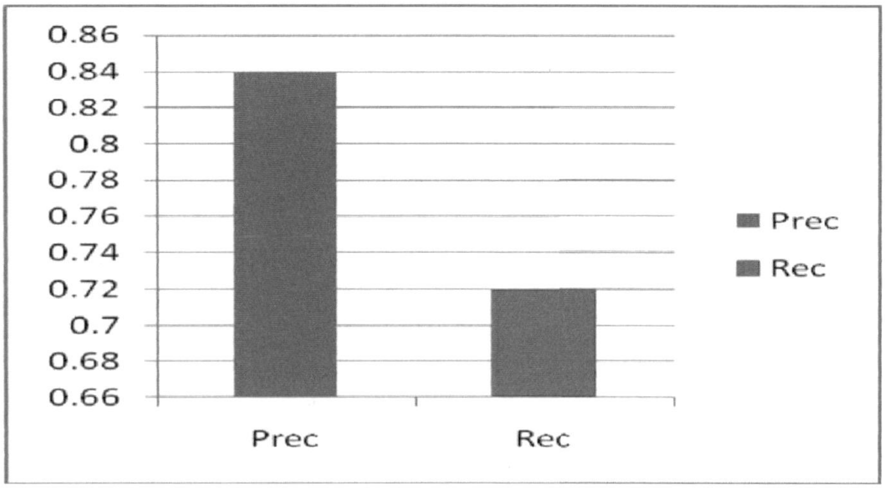

Figure.6 Propose Graph Precision and Recall

5. Conclusion

The main goal of this research is to show the importance of support vector machine in the effective retrieval of the picture in this SVM is used as the classifier which is performing task of classification the image and this process of classification are provide to all picture are evaluated after feature extraction process. CBIR, also known as the query through picture content material (QBIC) is the utility of computer imaginative and prescient strategies to picture retrieval problem, this is, the trouble of looking for virtual pictures in big databases. It aims to locate pictures of the hobby from a large picture database the usage of the visible content of the picture.

Reference

[1] C. Town, D. Sinclair, "Content-Based Image Retrieval Using Semantic Visual Categories", Technical Report TR2000-14, AT&T Labs Cambridge, 2000.
[2] R. Krishna, "Content-Based Image Retrieval Systems in Medical Applications - Clinical Benefits and Future Directions", International Journal of Computer Science & Network Security, Vol. 12, No. 9, p. 113, 2012.
[3] R.S. Kumar, M.S. Murugan, "Content-Based Image Retrieval System in Medical Applications", International Journal of Engineering, Vol. 2, Issue 3, 2013.
[4] K. Wanjale, T. Borawake, S. Chaudhari, "Content-Based Image Retrieval for Medical Images Techniques and Storage Methods - Review Paper", IJCA Journal, Vol. 1, No. 19, pp. 105-107, 2010.
[5] F. Long, D. Feng, "Fundamental of Content-Based Image Retrieval", Chapter 1, 2003 http://www.cse.iitd. ernet.in/~pkalra/siv864/Projects/ch01_Long_v40-proof. Pdf
[6] R. Rahimzadeh A. Farzan Y. Fanid Fathabad "A SURVEY ON SEMANTIC CONTENT BASED IMAGE RETRIEVAL AND CBIR SYSTEMS" International Journal on "Technical and Physical
[7] Problems of Engineering" (IJTPE), Iss. 18, Vol. 6, No. 1, Mar. 2014.
[8] Deepak Sharma, Deepti Sharma and Vikas Sindhu "An Approach to Digital Halftone Processing Using Error Diffusion in Forward and Backward Direction" Volume 3, Issue 6, June 2013 ISSN: 2277 128X International Journal of Advanced Research in Computer Science and Software Engineering.
[9] Prerana Karnik, Nitin Shahane "A survey on Content-based Image Retrieval using Vector Quantization" International Journal of Computer Applications (0975 – 8887) International Conference on Recent Trends in Engineering & Technology - 2013(ICRTET'2013).
[10] Behzad Merhrbakhsh Choobar, Saeed Mozaffari " A Robust Content Based Image Retrieval Using Local Full-Directional Pattern (LFDP)" 2017 IEEE.
[11] Abdolreza Rashno, Saeed Sadri "Content-based Image Retrieval with Color and Texture Features in Neutrosophic Domain" 3rd International Conference on Pattern Recognition and Image Analysis (IPRIA 2017) April 19-20, 2017.
[12] Devi S, Ajai Mathew "Fast Image Retrieval Using Error Diffusion Block Truncation Coding and Unsupervised Clustering" 2016 International Conference on Emerging Technological Trends [ICETT].
[13] Navneet Kaur, Sonika Jindal and Bhavneet Kaur "Relevance Feedback Based CBIR System Using SVM and Bayes Classifier" 2016 Second International Conference on Computational Intelligence & Communication Technology.
[14] Jing-Ming Guo, Heri Prasetyo, and Jen-Ho Chen "Content-Based Image Retrieval Using Error Diffusion Block Truncation Coding Features" 2014 IEEE.

PLC Operated Colour Based Product Sorting machine

S Vandana[1], K Shamukha Sri Sai[2], P Rohila[3] and V Manideep[4]

[1]Assistant Professor, Department of Electronics and Instrumentation, VNR Vignana Jyothi Institute of Engineering and Technology, Bachupally, Hyderabad
[2,3,4] Student, Department of Electronics and Instrumentation, VNR Vignana Jyothi Institute of Engineering and Technology, Bachupally, Hyderabad

[1]vandana_s@vnrvjiet.in

Abstract. Manual sorting requires excessive effort and takes time, while automated sorting makes it possible to sort items quickly and efficiently. Production and packaging systems widely use the sorting method of product lines as a bridge. Industries have been sorting processes to arrange the materials of their items according to the requirements to decrease labour and increase overall accuracy. We have developed a S71200 PLC-based sorting system to demonstrate this sorting system, where we use electro-magnetic actuators with TCS2300 colour sensor to automate the whole system. The system consists of a conveyor belt to carry items and a colour sensor to sense the product's colour form, and the system will then use 2 electromagnetic actuators to actuate and drive the product at its respective collection station. To identify the colour of the object moving on the conveyor belt, a colour sensor was used. The colour sensor transmits a particular signal to the PLC as soon as the appropriate colour is detected, which, in turn, operates the corresponding electromagnetic actuator to force the product out of the respective station. We can completely automate the sorting system in this way. In parallel processing industries, the sorting of these goods is very difficult. A consistency issue is created by continuous manual sorting. For controlling the overall operation, PLC is used as a master controller and Arduino as a slave controller. The colour identification method is based on the frequency analysis of the colour sensor's output. This will fulfil the requirements of higher output and accurate performance in the field of automation.

1. Introduction

Automatic sorting system is the fast and efficient system for sorting of products, whereas manual sorting is an effort intensive process. Industries use sorting systems to sort products/raw materials as per nominal criteria to improve accuracy and minimize the human effort. Manual arranging is a period extreme and exertion serious procedure. Programmed arranging frameworks consider snappy and proficient arranging of items. Our endeavour incorporates decreasing the cutting-edge cost used in the foundation and working of different vehicle line structures in different ventures, it is a fundamental negligible exertion automation system used for masterminding.

Several systems are proposed with automatic sorting of products in industries.In "Automatic Sorting in Process Industries using PLC".by P. Thirumurugan has proposed a system which sorts an object based on the height. But this system is implemented for only small ranges of height which cannot be used in large scale sectors [1].Mohd Aleem Uddin implemented “Automatic Industrial Sorting Machine by the different Sensors for Automation-An Innovative Model-Based Approach”. As per his examination, PLC can be viably intended for a wide assortment of control assignments with the simple ladder logic to control the whole framework. Be that as it may, this framework has detriment i.e, there is no visual investigation and moving of article and setting in just specific areas happen [2].

In "Design of an object sorting using PLC" the author proposed a structured framework that is completely mechanized utilizing PLC and automated DVD Drives additionally. The principle inconvenience utilizing of DVD drives is space occupation and they are costly contrasted with actuators and output activities fluctuated [3].In Object sorting and stacking automation using PLC a design that would count and segregate the works into the box, and then stack those things dynamically was proposed using Infrared sensors. Infrared sensors have disadvantages such as requiring line of sight, restricted range, and sluggish climatic conditions such as dust & fog which in reality might affect the sensors and data signal strength, therefore it might not be a sophisticated device [4].

In "PLC Based Object Sorting Automation" a setup that would significantly accomplish to a limited extent by using the low cost automation system (LAC) to avoid risk, improve repeatability, ramp up production speed also reducing the cycle time was developed. In industry, it can be used for sorting different objects, tools with a high degree of accuracy and automation quality [5].The author Saurin Sheth proposed "Automatic sorting machine using machine vision" in which an arranging framework that can check the flawed parts and put them in the reusing compartment. In this work, the imperfect items are distinguished and are set in the reusing receptacles and the costs of these machines are generally high so it is hard to use in little ventures and partnerships [6].

In "Automatic Sorting Machine Using Conveyor Belt" an ongoing application that helps in objects arranging with the use of PLC was developed. A few sensors are utilized to screen the parameters, for example, shape, tallness, weight, shading, and so on [7].A sorting system using PLC and HMI system was developed with automatic conveyer system with in-process sorting mechanism. The automatic conveyer and HMI systems can be used in wide range. But this system has many disadvantages. It is costly for small industries and it requires frequent monitoring and the PLC devices are proprietary [8].

In "Development of a Prototype Automated Sorting System for Plastic Recycling" an automatic sorting system for recycling of plastics was developed. The refinement of the routines in pattern recognition involves artificial intelligence techniques also in edge detection useful for high speed conveyer belts which can blur the acquired images was developed [9]. "A proposed hierarchical control model for automated manufacturing systems", describing the five layer hierarchical production control model was proposed to manage factories. The hierarchical structures need frequent changes and can lead to communication barriers [10].

The author Pourdarbani proposed a "Study on an automatic sorting system for Date fruits" in a journal of the Saudi Society of Agricultural Sciences. This system is developed based on Machine vision to sort the date fruits in food packaging industry. Main disadvantage of using this method can vary in the presence of outlier [11].

So, a novel automatic sorting system using actuators with a colour sensor which are controlled by PLC is developed. The sorting system consists of a conveyor belt to carry products from one end to the collection bins. This is designed with a proximity sensor and a colour sensor attached to the conveyor belt setup wherein which the proximity sensor distinguishes if the placed product is a metal or a non-metal and the colour sensor is for colour assortment.

2. Methodology

The main objective is to plan and build up a PLC worked colour-based object sorting machine, which is utilized to sort metals and non-metals followed by colour assortment. The output from TCS 2300 is less than required, which is only about 10% to 15% of the frequency. Due to this the information required cannot be processed from low output directly, for in which an Arduino is used to process the information and to obtain the output which is to be understood by the PLC controller.

The system utilizes three electromagnetic actuators to regulate operations. One actuator is used for metal/non-metal sorting and the other two actuators are used to sort two colours and for an already sorted product. The whole system is controlled by a PLC to regulate the sorting process. Three assortment bins are used to retain the products sorted by the actuators mounted parallel to it. The PLC communicate with sensors and actuators to attain the specified functionality and control total automated product sorting system.

For the most part, enterprises deliver similar models with almost no variation in dimensions, weight, shape and in this way arranging of items assumes a significant job here. In the days of yore, it was feasible to execute physical work for arranging comparable items. In any case, these days to limit the work use for such untalented errands, ventures can't manage the cost of human mistakes for arranging these items. This constrained enterprises to tend towards auto missing the arranging procedure. As the economy has consistently been an extensive factor in the creating business, in this way it got important to grow minimal effort computerization for precisely arranging these items.

2.1 Process flow & Block Diagram

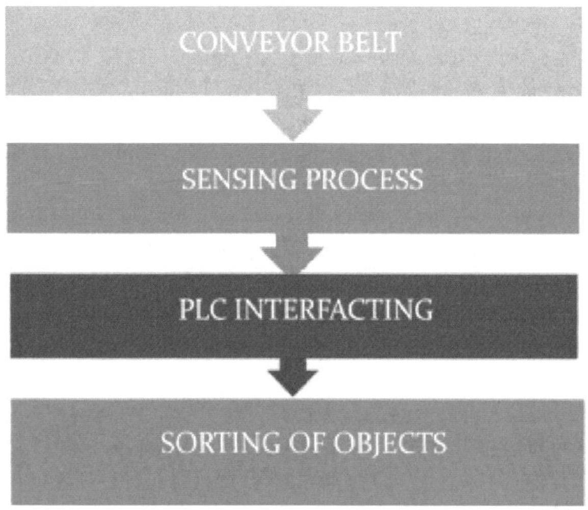

Figure 2.1 Process flow

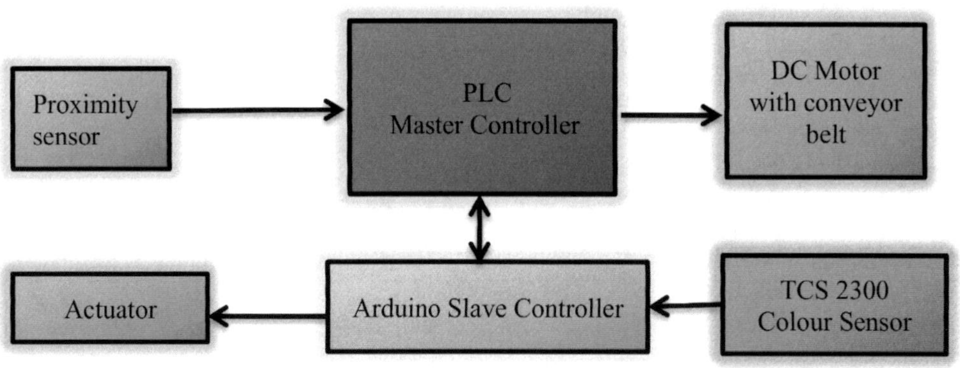

Figure 2.2 Block Diagram

2.2 Working of PLC based sorting system

The system that we have come up with aims to reduce human efforts by replacing manual sorting of things with automation and for fast and efficient sorting of products. This system is designed to sort two types of products. The products are placed on the conveyor belt, which moves it forward, the conveyor line is driven by 100 rpm motor of the DC type. The objects are moved forward for sensing, The DC motor is used to move the continuous movement of the conveyor belt. A relay, an electrically operated switch, used for the process of switching. Initially, the objects that are to be assorted are placed on the conveyor. The conveyor is started, and the object will start to move on the conveyor, then the object will pass through the sensor.

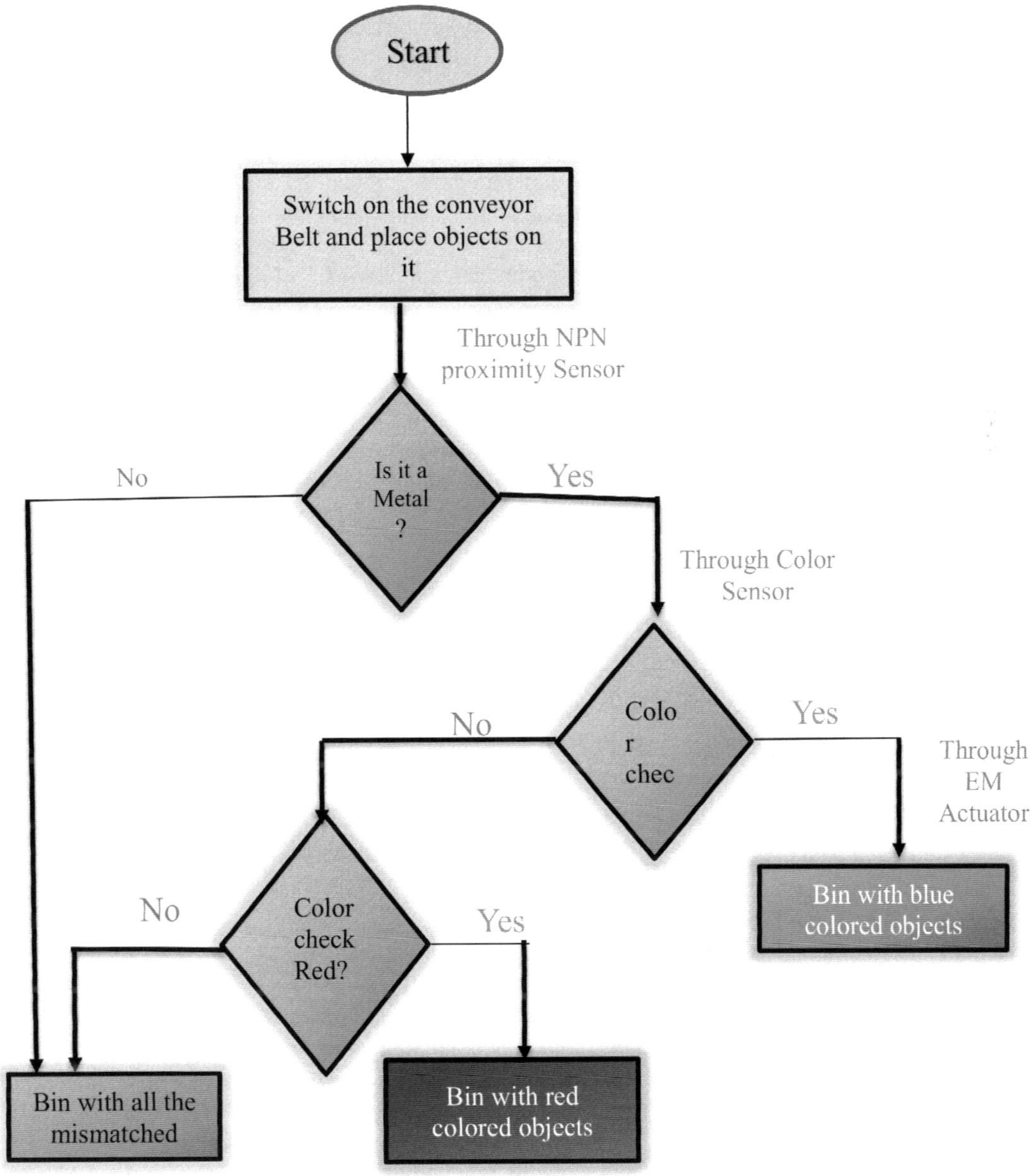

Figure 2.3 Flow Chart

The Proximity sensor senses the object coming parallel to it thus segregates metals and non-metals. It utilizes the inductive effect of the detection of metals. If the object is failed to sense near the Proximity sensor, then the product further moves on the conveyor belt and is sensed by the colour sensor. The colour sensor is placed to sense the objects with colour and then collected in the sorting bins. If the colour of the object is identified by the sensor then the object goes to forting sorting procedures as per requirement by going through the whole process and get sorted into waste bins.

The sorted bins are placed to collect the products after they are been sensed by the sensors. For this purpose, the Electromagnetic actuators are placed to push the products into the sorting bins. We have two actuators for this purpose. One of the actuators is used for pushing the products that are sensed by the proximity sensor. And the other actuator is used to push the products that are sensed by the colour sensor. This way various products are differentiated and categorized accordingly.

2.3 Flow chart description

- The packages are placed on conveyor belt line initially. The belt on the conveyor starts moving as a DC motor is connected to it.
- As soon as the object reaches to a position such that it is parallel to the proximity sensor, it is detected for its metallic/non-metallic nature. The response generated is sent to the PLC.
- After the sorting based on this characteristic the objects are then passed by parallel to the COLOUR sensor and used to sense the products of particular COLOUR if it matches it is sorted using actuators it goes to waste bins.
- In this way the objects are passed by the sensors and detected or sorted based on the requirements that we give in the ladder logic.

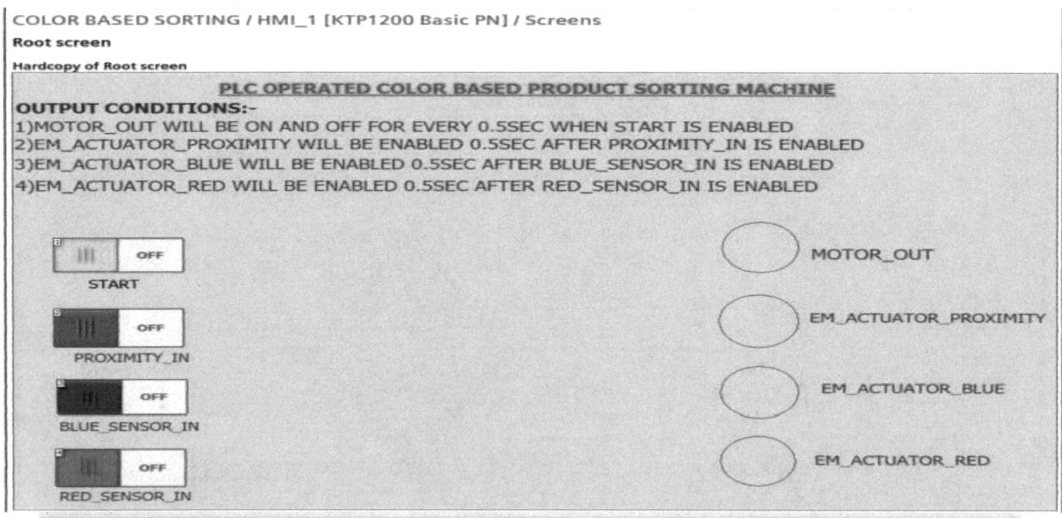

Default tag table

Name ▲	PLC tag	Address	Access mode	Acquisition cycle
BLUE_COLOR_SENSOR_IN	BLUE_OUTPUT	%M0.5	<absolute access>	100 ms
EM_ACTUATOR_BLUE	EM_ACTUATOR_2	%M0.6	<absolute access>	100 ms
EM_ACTUATOR_PROXIMITY	EM_ACTUATOR_1	%M0.4	<absolute access>	100 ms
EM_ACTUATOR_RED	EM_ACTUATOR_3	%M1.0	<absolute access>	100 ms
OUTPUT	MOTOR_OFF_CONDITION	%M0.1	<absolute access>	100 ms
PROXIMITY_IN	PROXIMITY_OUT	%M0.3	<absolute access>	100 ms
RED_COLOR_SENSOR_IN	RED_OUTPUT	%M0.7	<absolute access>	100 ms
START	START	%M0.0	<absolute access>	100 ms

Figure 2.4 HMI screen with tags

3. Results and Discussions

We made an ease programmed arranging framework utilizing the Programmable Logic Controller (PLC). We have attempted to arrange that will diminish human exertion and prevailing to a degree by utilizing the minimal effort robotization framework (LAC) to stay away from hazard, improve exactness,

IOP Conf. Series: Materials Science and Engineering 1119 (2021) 012016 doi:10.1088/1757-899X/1119/1/012016

and speed up creation. The arranging framework is utilized to sort the items dependent on their metals/non-metal attributes and shading put together qualities which are going with respect to the transport framework utilizing electromagnetic actuators. Further a few upgrades should be possible in our work by actualizing it in enormous scope areas. The arranging procedure should be possible from multiple points of view by arranging dependent on the different measures of the item, by size of the article or by dependent on the material by which it has been made. The procedure can be additionally improved by executing various kinds of arranging framework in an equivalent arranging procedure and it very well may be made with different sorts of arranging framework. Additionally, this can be additionally improved to an arranging framework that sorts the things dependent on the other physical thought. What is more, this can be accomplished utilizing the different sensors. In industry it tends to be utilized for arranging of different items, with high level of exactness and quality with a robotization.

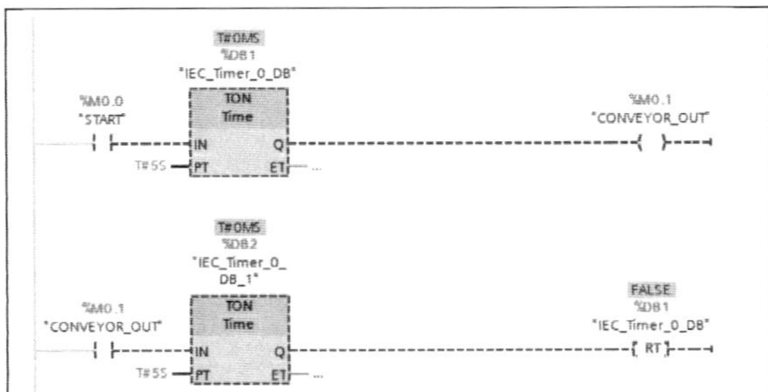	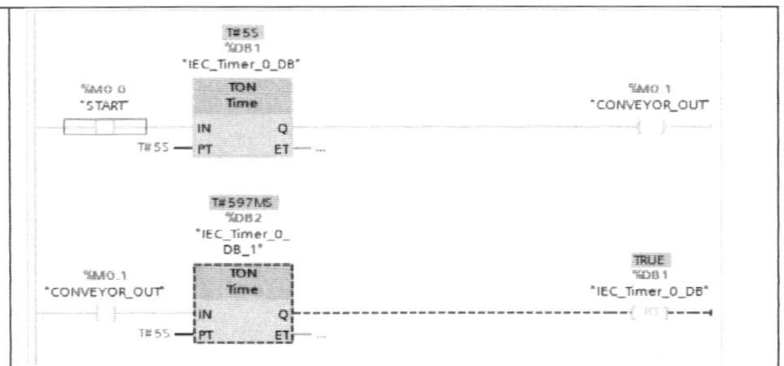
Conveyor will be in off condition when start is not enabled	Conveyor will be on & off for every 0.5sec when start is enabled
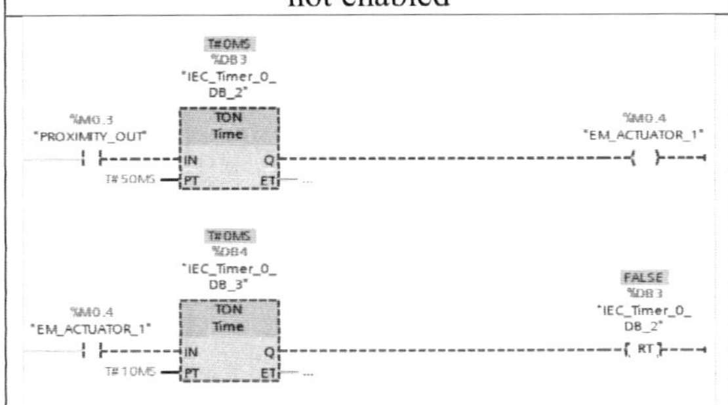	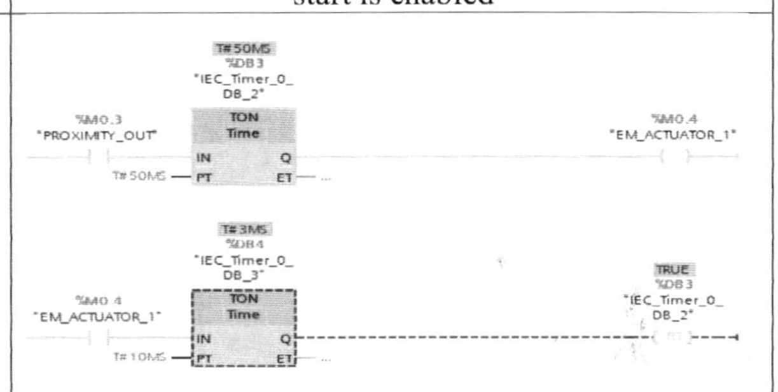
When proximity sensor output is null em_actuator_1 will be in off condition only	When proximity sensor output is enabled after 0.5sec em_actuator_1 will be enabled
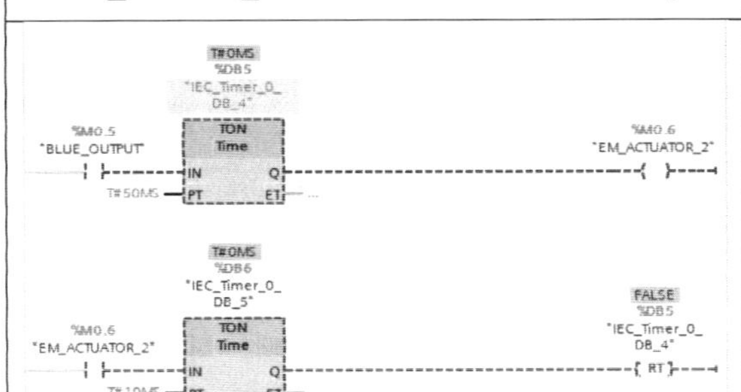	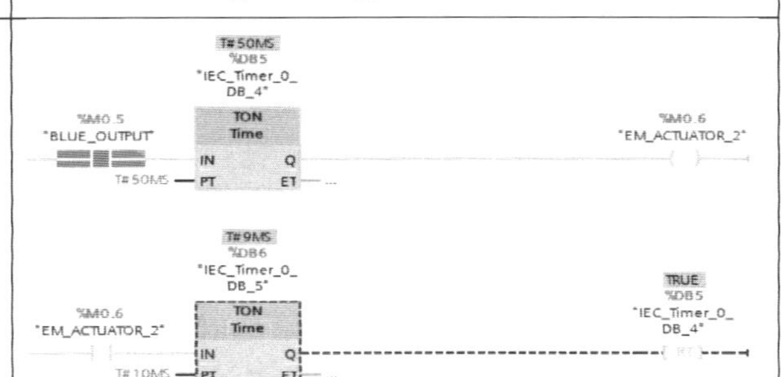
When blue colour sensor output is null em_actuator_2 will be in off condition only	When blue colour sensor output is enabled after 0.5sec em_actuator_2 will be enabled

IOP Conf. Series: Materials Science and Engineering 1119 (2021) 012016 doi:10.1088/1757-899X/1119/1/012016

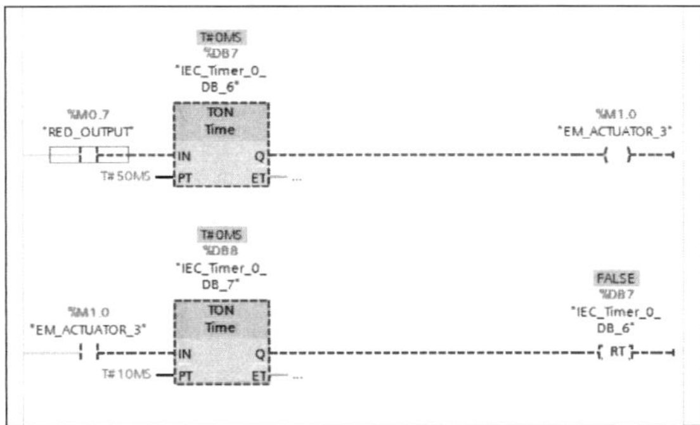

When red colour sensor output is null em_actuator_3 will be in off condition only

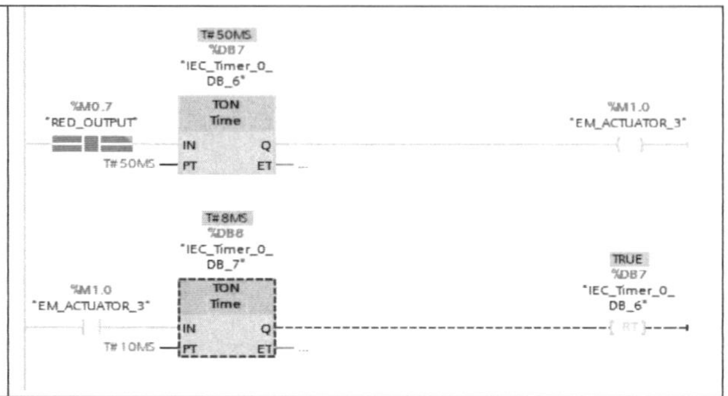

When red colour sensor output is enabled after 0.5sec em_actuator_3 will be enabled

Fig 3.1 : Ladder Logic Output

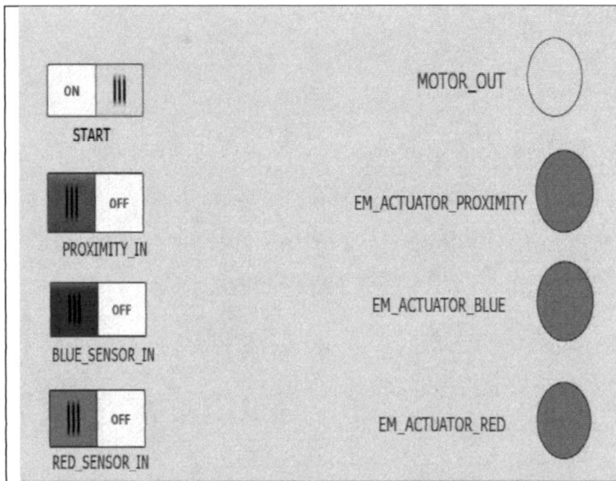

HMI screen when conveyor is in on condition

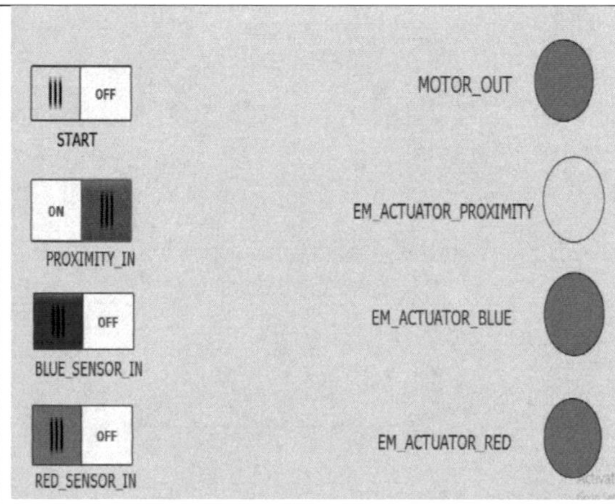

HMI screen 5sec after proximity_in is enabled

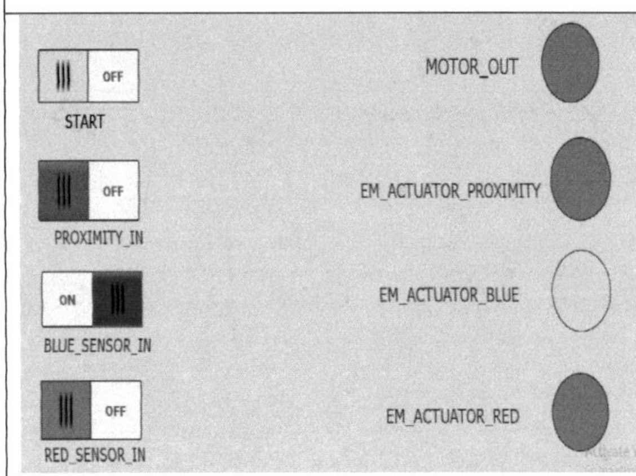

HMI screen 5sec after blue, sensor_in is enabled

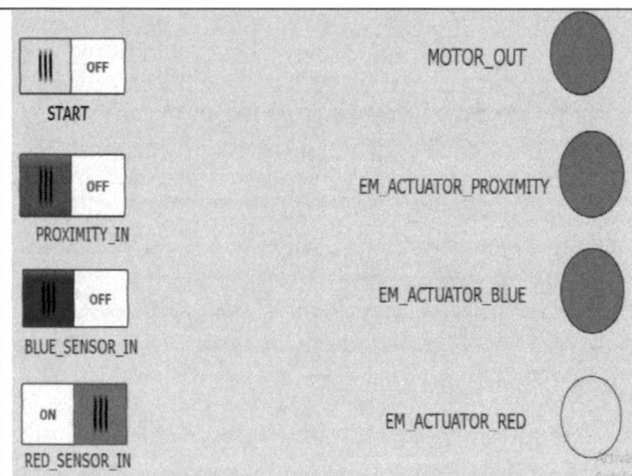

HMI screen 5sec after red sensor_in is enabled

Fig 3.2: HMI Screen Output

4. Conclusions

An easily reprogrammable framework utilizing the Programmable Logic Controller (PLC) was developed. An attempt was done to diminish human exertion and prevailing to a degree by utilizing the minimal effort computerization framework (LAC) to keep away from chance, improve precision, and speed up creation. The arranging framework is utilized to sort the articles dependent on their metals/non-metal qualities and shading put together attributes which are going with respect to the transport framework utilizing electromagnetic actuators. Further a few upgrades should be possible in our work by executing it in huge scope segments. The arranging procedure should be possible from multiple points of view by arranging dependent on the different physical parameters of the object like degree of size of or dependent on the type of material it is built. Furthermore, sorting can be accomplished utilizing the different sensors. In parallel process industry tends to be utilized for arranging of different items, with high level of precision and quality with a mechanization.

By developing this PLC operated colour-based object sorting system, sorting the objects based on their metal/ non-metal characteristics and based on their colour assortment was done. This sort technique will be exceptionally helpful in parallel processing industries where different objects are processed simultaneously. This differentiating procedure is a unique procedure by which we can sort n number of objects passing through conveyor in a single stretch without any errors. The sorting procedure additionally decreases the manpower requirement in numerous enterprises.

5. Future scope

Programmed arranging machine upgrades proficiency, common sense, and security of administrators. It guarantees wonderful preparing limit just as flawless execution including shading discovery. Of course, we have to include stepper motors or high precision motors and sensors with considerable reaction to accelerate the framework for mechanical sorting process. The present prototype can be developed further by rolling out certain improvements in the program and segments. A few proposals for developing this same work have been listed below.

- We can add other parameters in the sorting process by using different sensors which can make sorting process more efficient like weight, height and type of material of the object.
- By using DC motor, increases operation and maintenance cost due to presence of commutator and brush gear. So using stepper motor can make it more efficient.
- We can also track how many objects are being sorted by developing automatic counter which can count number of objects being sorted.
- Sensors with high precision and accuracy can be used to sort the object based on the colour assortment more accurately
- Usage of quality sensors make sorting more efficient and need not to be checked or replaced more often.

References

[1] "Automatic Sorting in Process Industries using PLC" by P. Thirumurugan GRD Journals- Global Research and Development Journal for Engineering | Volume 3 | Issue 3 | February 2018
[2] Mohd Aleem Uddin "Postal Automation System for Mail Sorting" International Journal of Emerging Technology and Advanced Engineering, Volume 5, Issue 3 December 2017.
[3] Shaik Faiz Hussain, Syed Ali Safdar, Mohd Aleem Uddin, "Design and Development of a PLC Based Automatic Object Sorting", International Journal of Research and Scientific Innovation (IJRSI) Volume IV, Issue XII, December 2017.
[4] Dhaval Tailor, Vivek Kamani, Ankit Ghetiya, Naresh Bhatiya, "Object Sorting and Stacking Automation with PLC", International Journal of Engineering and Technology (IJET) DOI: 10.21817/ijet/2017/v9i3/170903S031 Vol 9 No 3S July 2017.
[5] Prof. Nilima Bargal1, Aditya Deshpande2, Rucha Kulkarni3, Rucha Moghe4,"PLC Based Object

Sorting Automation"International Research Journal of Engineering and Technology (IRJET)Volume: 03 Issue: 07 | July-2016

[6] Saurin Sheth,Rahul Kher, Rushabh Shah, Parth Dudhat, Pratyush Jani "Automatic Sorting System Using Machine vision", DOI: 10.13140/2.1.1432.1448 Conference: Multi Disciplinary International Symposium on Control, Automation & Robotics, At DDIT, Nadiad, Volume:1.

[7] Y V Aruna, Beena S "Automatic convey or System with In–Process Sorting Mechanism using PLC and HMI System", Int. Journal of Engineering Research and Applications ISSN: 2248-9622, Vol. 5, Issue 11, (Part -3) November 2015, pp.37-42

[8] S.V.Rautu, "sorting of objects based on colour, weight, and type on a conveyor line using PLC, "IOSR Journal of mechanical and civil engineering (IOSR-JMCE), ISSN2278-1684.

[9] Darshil ; Sagar ; Rajiv ; Satyajit A. Pangaokar ; V.K. Sharma " Development of a PLC Based Elevator System with Colour Sensing Capabilities for Material Handling in Industrial Plant" Published in: 2008 Joint International Conference on Power System Technology and IEEE Power India Conference.

[10] D. A. Wahab, A. Hussain, E. Scavino, M.M. Mustafa and H. Basri "Development of a Prototype Automated Sorting System for Plastic Recycling" American Journal of Applied Sciences 3 (7): 1924-1928, 2006 ISSN 1546-9239 © 2006 Science Publications.

[11] Albert T. Jones, Charles R. McLean, "A proposed hierarchical control model for automated manufacturing systems", National Bureau of Standards, Gaithersburg, Maryland, USA

[12] Pourdarbani, R., et al., Study on an automatic sorting system for Date fruits. Journal of the Saudi Society of Agricultural Sciences, 2015. 14(1): p. 8

IOP Conf. Series: Materials Science and Engineering 1119 (2021) 012017 doi:10.1088/1757-899X/1119/1/012017

A Comparative Analysis of using Various Machine learning Techniques based on Drowsy Driver Detection

A Bano[1], A Saxena[2], G K Das[3]

[1] Department of Computer Science & Engineering, Compucom Institute of Technology and Management, Jaipur, India

[2] Professor, Department of Computer Science & Engineering, Compucom Institute of Technology and Management, Jaipur, India

[3] Assistant Professor, Department of Computer Science & Engineering, Compucom Institute of Technology and Management, Jaipur, India
E-mail: [1]arjuqureshi2020@gmail.com

Abstract: In image processing or computer vision, image segmentation is a vital issue for applications such as scene understanding, medical image evaluation, robotic perception, video surveillance, increased reality or compression, etc. Every year in road accidents caused because of human mistake, the numbers of dead and injured are rising. Drowsiness and driving are particularly risky and difficult to recognize. The second leading cause of road crashes in drowsiness after alcohol. Detecting driver drowsiness is a technology of safety for vehicles that helps placed an end to driver injuries that are dozy. One of the main causes of road accidents is driver drowsiness. It is a very serious issue for road safety. We have presented various methods for detecting the drowsiness of the driverin this paper and the comparisons among such methods are extremely challenging. For this purpose, we have compared machine learning methods based on facial expression,especially on eye state. Apart from eye detection, it performed experiments on mouth detection and face detection as well. This paper explores several methods for machine learning, like SVM, CNN, or HMM. From the analysis, we have found that the HMM model achieved more accurate results in comparisonto others.

1.Introduction

Drowsiness or fatigue is one of the key risks to road safety which causes serious injury, death, and expense. The increased drowsinessworsens driving performance. The unconscious shift from waking to sleep leads to a lack of alerting which leads to many major road injuries. Drowsy driving has resultedin over 100,000 road injuries and over 1,500 death rates each year, documented in U.S. National Highway Traffic Safety Administration (NHTSA). Fatigue of driver can have many factors, including sleep loss, long journey, restlessness, alcohol consumption & mental pressure. Each will lead to a severe disaster. Today, road rage has been in many parts of the past, causing drivers strain. The previous transport system is therefore not adequate to deal with these threats on the roads. Therefore, most deadly injuries can be avoided by embedding automatic fatigue detection systems in vehicles. The drowsiness detection system (DDS) analyzes drivers' attention and alerts the drivers before reaching the serious road safety threat [1].

In our life, emotions play a crucial role. During our lives and interactions with others, there are many types of emotions. Recent research is done on emotion recognition. If this is applied in the social networking setting, it can be an important way to understand how individuals and social circles, neighborhoods, and towns feel about recent or different events [2,3].

Machine learning has become one of the major stays of IT. There is a strong purpose to assume that intelligent data processing will become much more widespread as an essential factor for technical purposes, considering the growing amount of information available. Most of the science of ML is about solving these needs and providing the solution [4].

The organization of this paper as follows Section II introduces image segmentation including image classification techniques. In Section III we describe the drowsy driver detection system in this paper. We present a detailed description of numerous Computer vision methods to detect driver drowsiness in Section V. Next Machine learning techniques are described in Section VI. After this, we provide a detailed study of related work. Conclusions are presented in the last section.

2.Classification of Segmentation Algorithms

Segmentation algorithms havebeen created to segment imagesfrom segmentation; its dependent on 2 key characteristics, discontinuity & similarity. Division and subdivision are predicated on discontinuity depending on the abrupt intensity or gray picture levels. Our interest in this approach is primarily to define individual points, lines, and boundaries. The predicted pixel community in heterogeneous attributes contains approaches such as thresholds, regional expansion, and region-splitting as well as merging.

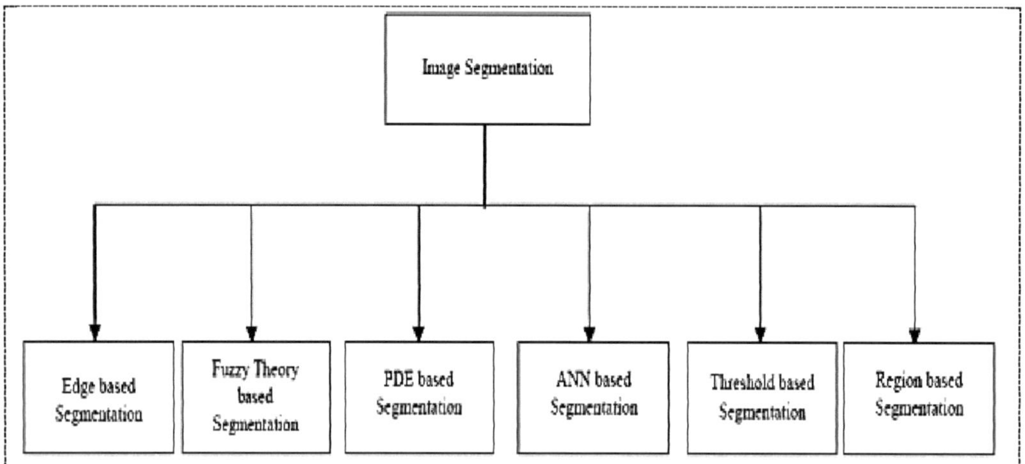

Figure 1. Image segmentation Methods.

2.1. Segmentation by Edge Detection

Edge detection is a fundamental step in the segmentation process. The image splits into the object as well as its context. Divide the image by the edge detectionprocess to detecting the change in intensity or image pixels. Two key methods for edge detection embedded in the segmentation are found in the Gray histogram and Gradient. The canny edge detector is a second-order derivative operator that is split into two types of edge detection operators, as first-order derivative & second-order derivative operators. The operators with a second-order derivative provide reliable results.

2.2. Segmentation by Thresholding

Image segmentation is one of the easiest methods for the image which is dependent on levels of intensity named as a threshold. Thresholding can be implemented either globally or locally. Global threshold differentiates object & context pixels by comparing them to the selected threshold value and by using a binary partition in the image section. An oftencalled adaptive threshold is the local thresholding strategy. Adaptive thresholding strategyis threshold value varies according to the locale characteristics of the regions in the image and histogram thresholding is usedfor the segmentation of the specified image; some preprocessing & post-processing approaches are essential for threshold segments. Mean technique, P-style method, histogram-based method, Edge maximization technique & visual method are primary thresholding techniques suggested by various researchers.

2.3. Segmentation by Region-based
The segmentation region observed should be closed. Also, the region-based segmentation is called the similarity primary based segmentation. Due to the lack of edge pixels, there is no difference in the region primarily based on segmentation the boundaries are known for segmentation. The sting flow is regenerated into a vector after distinctive alteration in the color and texture. The sides are detected for further segmentation.

2.4. Segmentation by Feature
- Based Clustering
- Segmenting by K Means Clustering[5]

3. Driver Drowsiness
Drowsiness is simply referred to as "fatigue-related near-sleep." It's technically distinct from fatigue, distinct as a “refusal to carry out the task at hand”. There are the same effects of sleepiness and fatigue. Fatigue has an effect on mental alertness, reduces the ability of the person to drive a vehicle safely & raises the risk of human error, which can lead to damage and injury. Sleepiness slows down the time of response, decreases awareness, and undermines judgment. Both transport operators, such as pilots, truck drivers, and railway engineers, have been impacted byfatigue and sleep deprivation. In these cases, the driver cannot focus on the main driving task, which will increase the risk of an accident. This problem would deteriorate further with ever-growing traffic conditions [6].

3.1. Drowsy Driver
Drowsy driving is the riskiest groupings of sleepiness or fatigue. Usually, driving sleep deprivation has become a growing risk of an accident; because the driver was not properly asleep, often owing to sleep syndromes, alcohol effects, few medicines, and even day & night activities.
If the driver is drowsy it increases the risk of a collision more often than normal driving. Nowadays it is very difficult to know the exact moment when sleep comes over. Also, the risk is raised during the night.

3.2. Factors Cause Driving Drowsiness
Drowsiness driver affects driving greatly and raises the risk of crashes. More than 30% of driver sleepiness or fatigue is responsible for road accidents.There are 2 key reasons for driver fatigue:
- If sleep quality &quantity are insufficient.
- Start driving when sleeping is usually finished [7].

3.3. Drowsy Driver Detection System
Many different methods can be used to assess the drowsiness of the driver. These methods are image recognition-based techniques, Electroencephalograph techniques, and techniques based on artificial neural networks. Image processing techniquesmay be split intothree types. These are the technique of matching the template, the technique of eye blinking and yawning. These methods are focused on the interpretation of images through computer vision. In computer vision technology the driver uses facial features e.g. blindness of the eyes and head movements to detect driver drowsiness.

Table 1. Techniques for drowsy detection

No.	Technique used	Strongpoint	Weak point
1.	Brain-computer interface [8]	Very efficient in health care	Electrodes outside of the skull can detect very few electric signals from the brain
2.	Geometry based [9]	Small database, recognition rate 95%	A large number of features are used.
3.	Template-based [10]	Recognition rate 100%	Complex
4.	Color-based [11]	Simple and small database	Limited performance
5.	Support vector machine [12]	Flexible	Lack of transparency in result
6.	Iris recognition [13]	Produce high accuracy result in less time	Expensive

3.3.1. Image processing-based Techniques
Drivers face images are used for the analysis of IP techniques so that their states are detected. From the face image, one can see that driver is awake or sleeping. The driver is asleep or drowsiness with the same images because the eyes of the driver are closed in the face image. And the facial image can also detect other signs of drowsiness. These strategies can be divided into three sub-categories:

1) Eye Blinking Based Technique
The blink rate andduration of the eye closure are measured to detect the drowsiness of the driver. Since eye blinkers and gaze between the eyelids differs from normal situations when the driver is sleepy at the time, so they detect drowsiness easily. The position of iris and eye conditions in this system is tracked over time to measure the duration of blink and the near life of the eye. And a remotely placed camera uses this sort of system to achieve video and computer vision techniques, such that visual, eyes, and eyes are sequentially located in the closure ratio. The driver's drowsiness can be identified by this eye closing and blinking ratio.

2) Template Matching Technique
In this technique, the eye state should be used for some particular timei.e. when the driver closes the eye/s then the system produces an alert. And, there is a driverthat has both open and closed eyestemplatein this system. It can also be used to open and close eye templates of the drivers.

3) PERCLOS Technique
PERCLOS is a fixed parameter for drowsiness detection level. The PERCLOS score is intended to determine if the driver is in a drowsy condition or not (the percentage of time an eye is closed at a certain time).

4) Yawning Based Technique
Yawn is drowsiness symptoms. yawn is ostensibly shown on a big vertical opening of the mouth. The mouth is wide openis wider than in the yawncompared to the speech. They can detect yawn by using face tracking and mouth tracking [14,15].

4.Computer Vision Method to Detect Driver Drowsiness

Computer vision techniques for identification of changes in facial expressions of the driver.

Computer vision approach to detect driver drowsiness using machine learning algorithms to detect the closure and opening of the eyes. There was no fully automatic model previously built. Second, a Low-cost solution has not been existingdriver drowsiness detection. Thirdly, no model worked for different luminance conditions (produced by sunlight & during dim light conditions similar to bad weather). Thus,there is an enormous need for a low costfor the fully automated driver drowsiness detection model, which operates in different luminance conditions.

4.1. Facial Expression

Three facial detection approaches are available. They are focused on features, templates, and appearance. The feature-based approach detects invariantface features but it is difficult to extract the featurein complex contexts. A predefined standard face pattern is used in the template-based approach and the correlation is used to identify the face. But it is difficult to expand different scales with this method. Face and non-face are observed in appearance-basedapproaches. But only with a simple background, this approach will produce reliable results. Facial expression characteristics are retrieved using the LG expression system. Six simple facial expressions are happiness, surprise, sadness, anger, hate, and fear.

4.2. Facial Tracking

The face tracking system must be robust for head movements, rotation, changes in illumination, and variation. For this purpose, a method has been suggestedfor the simultaneous use of face recognition & object tracking systems. This grouping provides us the chance to take advantage of 2 different programs together.

4.3. Eye Detection

It is difficult to find an eye because of various variables such as lighting, posture, facial shadow, etc. Different measures could be measured with the eyelid closure percentage, maximal closing time, blink frequency, average eye-opening level, eye-opening speed, and eye closing speeds, and an efficient driver drowsiness detecting model that can work under variedunconstraintandluminanceconditions.This condition can then be divided into distractions and the driver must alarm [6].

5. Machine Learning

Machine learning is a theoretical area that focuses systematically on theory, results, & properties of education systems & algorithms. It draws on several diverse areas of ideas: optimization theory, artificial intelligence, psychology, knowledge processing, optimal controls, cognitive sciences, also other branches in research, science, and mathematics. It is a strongly interdisciplinary area. The machine learning area is usually categorized into 3 sub-domains: supervised, unsupervised, & reinforcement learning [16].Let us discussed the various types of machine learning algorithms:

5.1. Supervised Learning

In supervised learning, we learn anobjective function that can be applied to predict the values of approved or not approved discrete class features. In a specific sample set, a machine learning algorithm makes predictions, whereas a supervised learning algorithm looks for designs to assign the data point labels. This algorithm is anoutcome variable from which a certain number of predictorscan be estimatedi.e. independent variables. Will create a function that maps input to desired outputs using this set of variables. The training phase continues until the model achieved accuracy in terms of training data. This whole procedure aims to reduce manual checks and coding expenses. Supervised learning examples are NNs, Regression, Decision Tree, NN, SVM, Naive Bayes, etc.

5.2. Unsupervised Learning

The learning useful structure is referred to as unsupervised learning with the labeled classes, criteria for the optimization, input signal, or other knowledge beyond the raw data. We don't have any target variable in this algorithm for calculating mean and we do not have a data points label or can say the training data class label is unknown here. This algorithm is applied to collect data into cluster group to explain the configuration of the data i.e. the cluster of the data which shows essential partitions and hierarchies. Examples: K-means, Clustering Fuzzy, Clustering Hierarchy Data are not labeled as well as the result is not known.

5.3. Reinforcement Learning

The machine is trained to make specific decisions using this algorithm. Based on each data point, these algorithms choose an action and later learn how to achievea successful decision. This is an environment in which the machine trains [17].

6. Literature Survey

S. Hachisuka (2013) used the active Appearance model (AAM) for the three-dimensional coordinates of the feature's points on the face images. They also used the k-Nearest-Neighbor approach to divide drowsiness into 6 levels. Consequently, among 13 participants, the average root means square error (RMSE) was below the 1.0 level. The first step in emotion detection was the smile and speaking detection[18].

B. N. Manu et al.(2016) discussed three well-defined steps efficient approach for detecting drowsiness. These three steps for facial detection are Via Viola-Jones, eye tracks, and bowel detection. The majority of skin's non-facing backgrounds are refused only by the chromatic components. The lighting may be invariant by segmenting the dermal component alone before the level is identified. By comparing the prototype with the correlation coefficient, eye tracking, and yawning has achieved [19].

J. Yan et al. (2016) built a grayscale image processing in a real-time drowsiness detection system and PERCLOS to operate if the driver is fatigued[20].

W. Hussein and M. S. A. El-Seoud(2017) Based on the combinations of the related functions, four SVM classification models havebeen developed. The study of these models shows that the maximum precision of the wavelet coefficients, form, circularity, and black ratio were obtained (91.3 percent). The findings from the methodology proposed demonstrated its promising inline usage in the cabin to assess driver drowsiness status [21].

R. Roopalakshmi et al. (2018) proposed a driver drowsiness detection system that uses blink counts to detect drowsiness. The experimental findings of the proposed method introduced with a single camera view in Open CV & Raspberry Pi settings have to demonstrate the system's good efficiency in accurate drowsiness detection and thus minimize road accidents. [22].

J. W. Baek et al. (2018) proposed for detects the face of the driver in this image and calculated the points of interest in the facialarea. To detect the face, an AdaBoost classifier based on Census Transformation features is added to the suggested algorithm. And for the face landmark recognition, the suggested algorithm used regressing local binary features. They got a dataset from the infrared camera, which uses real area, using video recordings. The proposed algorithm in the targeted board hasbeen tested (i.mx6q). Findings indicated that the proposed algorithm exceeded speed and accuracy [23].

B. Eraldo et al. (2019) Using Raspberry Pi3 in the OpenCV library, and sensors like MQ-3 that measures alcohol percentage and S9 that measures heart rate was created as part of a drowsy detection that integratedimage processing. It also has an alert system and a touch screen as an interface for displaying the data measured by the sensors. This method is therefore efficient and feasible [24].

R. K. M et al.(2019)implemented a system to detect drowsiness with the Raspberry Pi and including many sensors such as the Gas Sensor as well as the vibration sensor. A camera that captures the important sign is watched by the driver. The image of the individual is sent if the eyes are closed for a longer period. The accident is detected by a vibration sensor and notified with a latitude and longitude message to the server[25].

7. Problems Associate in Driver Drowsiness Detection

Security is our primary priority during travels or driving. A driver'smistake can lead to severe physical injury, death, or severe economic loss. Many systems, such as navigation systems, sensors, etc. are currently available on the market to make the driver work easy. There are many causes, in particular human mistakes that lead to road crashes. After recent years,sources show that there has been a massive increase in road injuries in our country. The main cause of traffic crashes is the driver's sleepiness and fatigue when driving. An effective technique is required to detect drowsiness as soon as the driver is sleepy. This will save a lot of accidents.

One of the major causes of road accidents is driver drowsiness. This is a major issue for road safety. If drivers have warned of these crashes before they were too drowsy to drive in safety. To detect drowsiness reliably, it relies on the timely warning of drowsiness. Drowsy is one of the main reasons behind traffic accidents which place the driver significantly at higher risk of accidents than driving alertly.

8. Measures and Techniques Used for Detection Drowsiness

Numerous facial features may be drawn from the face to deduce the degree of drowsiness. These compriseeye blinks, head movements & yawning. Nonetheless, it is a task to build a drowsy detection method that offers consistent & accurate results as detailed and robust algorithms are required. A variety of methods fordrowsy detection have been examined in the past time. The recent increase in deep comprehension requires revisions of such algorithms to detect drowsiness accurately. This paper explores techniques for machine learning that include SVM, CNN, and HMM in the context of the detection of drowsiness.Table 2 and the included are briefly described the approaches below.

8.1. Support Vector Machine (SVM)

SVMs have supervised classification and regression approaches to learning. Boser, Guyo&Vapnik initially introduced SVMs in 1992. SVMs aim to find a hyperplane that divides training data into predefined paths. In the field of driver drowsiness, SVMs are mainly applied to identify the various driver state from the defined data. There have been many efforts in the detection of drowsiness using the SVMs. A variety of measures were used to assess driver drowsiness rate using SVMs. Table 2 presents a comparison of these steps and briefly describes the methods below.

8.2. Hidden Markov Model (HMM)

HMMs are a mathematical model used to simulate hidden states based on detected probability-defined parameters. Leonard Baum and colleagues developed HMMs at the end of the 1960s and early 1970s. Today, HMMs are commonly used for applications like facial expressionrecognition, gene annotation, DNA sequence error modeling, and computer virus classification. Table 2 indicates different features and methods applied by HMM-DDs, but Zhang et al. and Choi et al. omitted details essential to compare their results. The Authors suggested a new facial feature using wrinkle changes that are detected by measuring the intensity of the local edge on the face. They have also applied an IR camera to avoid changes in light to allow including day-night operation. Fortunately, if peopleused this older system could give false results since the wrinkles are deeper. By contrast with HMM, methods for eye-tracking based on color and geometric features have been applied. Authors use a two-stage Lloyd-max calculation to extract light to ensure that changes in lighting are robust. This machine is unfortunately built for enclosed operating conditions & if the driver is not onward then fails to recognize the face.

8.3. CNN

CNN's are the same as a normal neural network and often consist of neurons for learning weights. CNN uses spatial layers that are suitable for images and that represent effective spatial correlations. CNN has proved effective in the areas of imagerecognition, video analysis, and classification. Yann LeCun and YoshuaBengio were the first to provide a computer vision in CNN. Table 2 presents a short review of the CNN-based drowsiness detection methods [26].

8.4. Viola-Jones Algorithm

Viola-Jones technique is based on input image processing or extracts pixel-based information to aid in the detection of features. Optimization of this window detects image faces of different sizes. It is an invariant detector in sizes also and runs through the image with different sizes each time. The required component issimilar no. of calculations, regardless of the image size, because it is invariant in scale. Viola-Jones is a cascade detector. The first phase contains detectors that only remove those image parts that do not face. The complexity of detectors has been further extended in later phases to further analyze the characteristics. A face is only detected as the entire cascade is passed through [27].

Table 2. SVM, HMM & CNN Techniques on Drowsiness Detection.

Author	Year	Metric	Methods	Classifiers	Accuracy%
F. Zhang, J. Su [28]	2017	Eye state	AdaBoost, LBF & PERCLOS	CNN	95.18
A. George and A. Routray [29]	2016	Eye state	Viola & Jones algo	CNN	98.32
A. Punitha, M. K. Geetha [30]	2014	Eye state	Adaboost& HMM	SVM	93.5
M. Sabet [31]	2012	Eye state	LBP-SVM	SVM	98.4
B. Zhang [32]	2012	Eye state	Cooccurrence matrix of oriented gradients (CMOG)	HMM	95.9
A. Bagci and R. Ansari [33]	2004	Eye state	Baum-Welch algo	HMM	99.7
Luo, R. C.[34]	2020	Eye state	Perspective-n-Point (PNP)	multi-model fusion	75.447%

Then,the SVM was trained to class as well as to trigger an alarm when the eyes open or closed. The developers of [30,31] suggested a method to detect drowsiness and distraction of the driver as well. So, this Viola & Jones algorithm was used with local binary patterns (LBP) to detect faces and color histograms to track faces over frames. The system produced an accuracy of 100% in face recognition, but the expected low frame rate that can lead to missing facial expressions which is potential decreases. The variety of features and approaches used by HMM drowsiness detectors are seen in Table 2 but Zhang et al. [32] omitted the details needed to compare its results and it is not part of the meta-analysis stage. Instead [33], introduced HMM eye detection methods based on color and geometric features. The authors used a 2-level Lloyd-max empirical to be stable for illumination changes [33]. This system is predictably conceived to be enclosed and fails to identify the face of the driver who doesn't look ahead. Table 2 also provides a brief overview of CNN & multi-modal fusion detection methods.

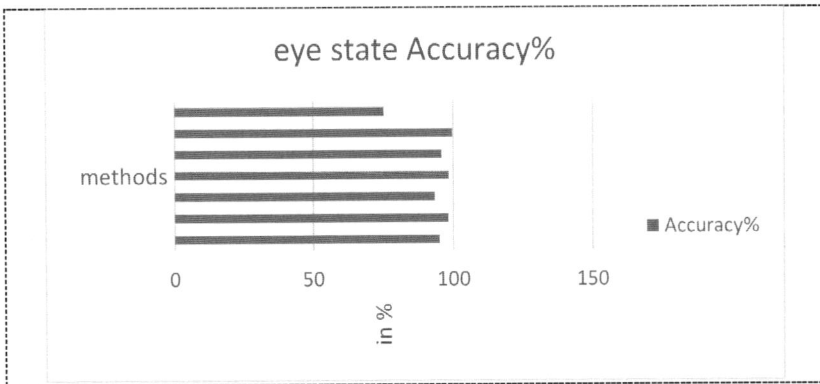

Figure 2. Comparison of all metrics.

As an outcome, it is difficult to compare methods by only estimating reported accuracies. An analysis of measures creating the driver's drowsiness detection system is now explored on machine learning techniques to identify various levels of drowsiness.

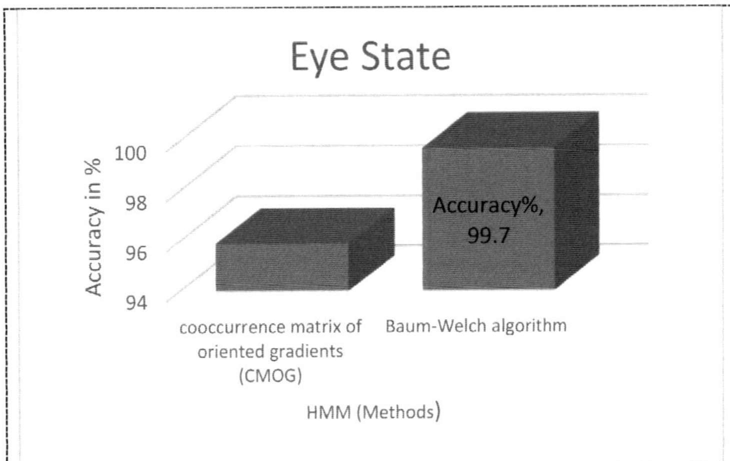

Figure 3. Accuracy of HMM for Eye State.

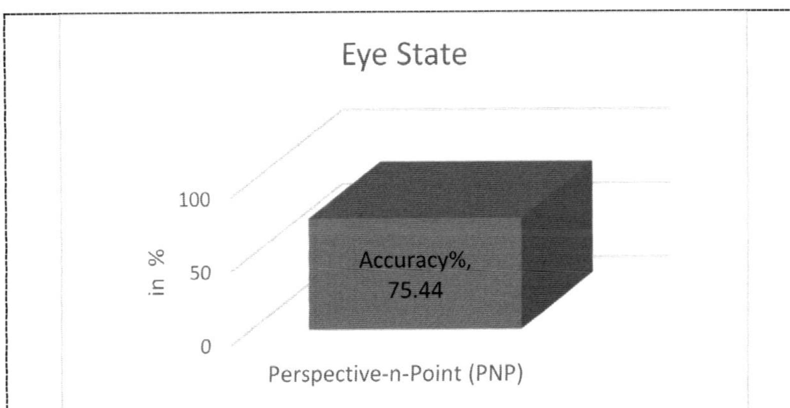

Figure 4. Accuracy of Multimodel Fusion for Eye State.

Table 3. Accuracy of eyes, Face, Mouth detection [35].

Metric	Total frame	Correct frame	Error frame	Miss frame	Accuracy
Face detection	3778	3758	3	17	99.47 %
Eyes detection	3761	3548	89	124	94.33 %
Mouth detection	3761	3664	97	0	99.80 %

Table 3 shows the accuracy rate for Face detection, Eyes detection, and Mouth detection metrics on the frame numbers.

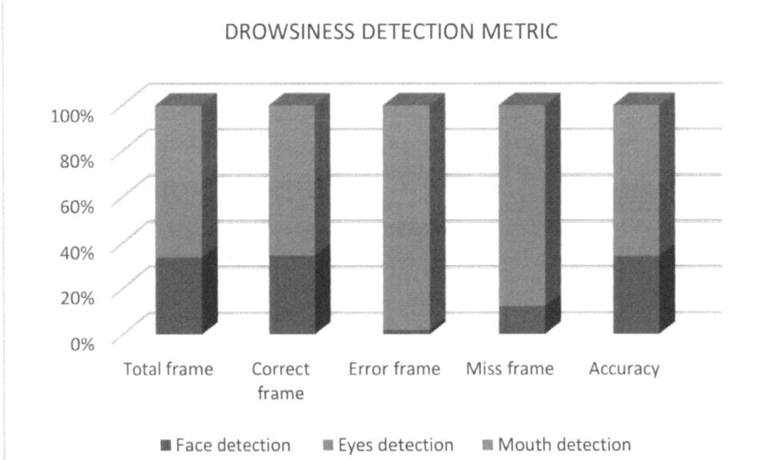

Figure 5. Comparison graph of Eyes, Face, and Mouth detection for Drowsiness Detection.

9.Conclusion

Many vision applications need precise and effective image segmentation methods and classification processes to analyze visual information & perform real-time decision making. Drowsy driving may be almost as fatal as drunk driving. The drivers' drowsiness doesn't only lay itself but it risks for everyone else. Tired & sleepy drivers have delayed reactions & have made poor decisions. One of the main advantages of the drowsy driver detection system is promotingsafety. Several technologies are obtainable to identify the drowsiness of the driver and every approach through its limitations. In this work, we have discussed and analyzed driver drowsiness with their approaches. There are various machine learning methods available to detect driver drowsiness. Machine learning is a paradigm that could apply to learning from previous experience to improve future success (in this case prior data). This paper consists of a support vector machine, convolution neural network, Hidden Markov model, and multi-model fusion classifiers. Such all classifiers usedseveral different methods and hybrid methods to detect drowsiness using eye state metric.It revealed that SVM technology is the most widely applied method to detect drowsiness, and recently CNN is the most suitable classifier, but HMM is performed better than the other two. It also tested face detection and mouth detection metrics with eye detection metric on the frame numbers and we found that drowsiness can detect more accurately (i.e. 99.80%) using mouth detection metric.

10. References

[1] Ramzan M, Khan H U, Awan S M, Ismail A, Ilyas M and Mahmood A 2019 *IEEE*7 pp 61904-61919
[2] Kumar K S 2020 *Int. Res. J. of Eng. and Technol. (IRJET)* **07**3
[3] DhargaveS C 2015 *Int. J. of Current Eng. and Scient. Res. (Ijcesr)***2** 2
[4] Kaur S and Jindal S 2016 *Int. J. of Innov. Res. in Adv. Eng. (IJIRAE)***3**11
[5] MariyammalC 2018 *Int. Adv. Res. J. in Sci. Eng. and Technol (IARJSET)***5**11
[6] Chisty 2015 *Int. J. of Comp. Sci. Tren. and Technol. (IJCST)***3** 4
[7] Kumari B M K and KumarP R 2017 *Int. Conf. on Big Data Anal. and Comput. Intell. (ICBDAC)*
[8] Dunbar J, Gilbert J E and Lewis B2020 Exploring differences between self-report and electrophysiological indices of drowsy driving: A usability examination of a personal brain-computer interface device *J. of Safety Research***74**pp 27-34
[9] Farahmand B and BoroujerdianA M 2018 Effect of road geometry on driver fatigue in monotonous environments: A simulator study *Transportation Research Part F: Traffic Psychology and Behaviour***58**pp 640-651
[10] Falahpour M, Chang C, Wong C W, Liu T T 2018 Template-based prediction of vigilance fluctuations in resting-state fMRI *NeuroImage***174**pp 317-327
[11] Panning A, Al-Hamadi A and Michaelis B2011 A color-based approach for eye blink detection in image sequences*IEEE International Conference on Signal and Image Processing Applications (ICSIPA)* pp 40-45
[12] Hu S and Zheng G 2008 Driver drowsiness detection with eyelid related parameters by Support Vector Machine *Expert Systems with Applications* **36**4 pp 7651-7658.
[13] Perez C A, LazcanoV A and Estevez P A 2007 Real-Time Iris Detection on Coronal-Axis-Rotated Faces I*n IEEE Transactions on Systems, Man, and Cybernetics, Part C (Applications and Reviews)***37**5 pp 971-978
[14] Nair I R 2016 *Int. J. of Eng. and Comp. Sci.***5** 11 pp. 19237-19240
[15] Choudhary P 2016 *Int. Res. J. of Eng. and Technol. (IRJET)* **03** 12
[16] QiuJ et al. 2016 *EURASIP J. on Adv. in Sign. Pro.*
[17] Kaur S and Jindal S 2016 *Int. J. of Innova. Res. in Adv. Eng. (IJIRAE)***3** 11
[18] Hachisuka S 2013 *Int. Conf. on Biometrics and Kansei Eng.* pp 320-326
[19] Manu B N 2016 *12th Int. Conf. on Innova. in Info. Tech. (IIT)* pp 1-4
[20] Yan J, Kuo H, Lin Y and Liao T 2016 *Int. Sympos. on Comp. Consu. and Cont. (IS3C)* pp 243-246
[21] Hussein W and El-SeoudM S A 2017 *Euro. Conf. on Elec. Eng. and Comp. Sci. (EECS)* pp 261-265
[22] Fouzia, Roopalakshmi R, Rathod J A, Shetty A S and SupriyaK 2018 *Second Int. Conf. on Invent. Comm. and Comp. Technol. (ICICCT)* pp 1344-1347
[23] Baek J W, Han B, Kim K, Chung Y and Lee S 2018 *Tenth Int. Conf. on Ubiqu. and Fut. Net. (ICUFN)* pp 73-75
[24] Eraldo B, Quispe G, Chavez-Arias H, Raymundo-Ibañez C and Dominguez F 2019 *IEEE 39th Cent. Amer. and Panama Conv. (CONCAPAN XXXIX)* pp 1-6
[25] R K M, R V, and Franklin R G 2019 *Int. Conf. on Vis. towards Emer. Tren. In Comm. and Net. (ViTECoN)* pp 1-5
[26] Ngxande M, Tapamo JR and Burke M 2017 *Patt. Recog. Ass. of South Africa and Robo. and Mech. (PRASA-RobMech)*
[27] BangalS, Bangal S and Sagvekar V2017 *Int. J. of Engg. Res. & Technol. (IJERT)*5 1
[28] Zhang F, Su J, Geng L and Z. Xiao Z 2017 *Proc. Int. Conf. Mach. Vis. Inf. Technol. C.* pp 105–110
[29] George A and RoutrayA 2016 *Int. Conf. Sign. Process. Comm.* pp 1–5
[30] Punitha A, Geetha M K and SivaprakashA 2014 *Int. Conf. Cict., Power Comp. Technol. (ICCPCT-2014)* pp 1405–1408

[31] Sabet M, Zoroofi R A, Haghighi K S and SabbaghianM 2012 *20th Iran. Conf. Elec. Eng. (ICEE-2012)*pp 1247– 1251
[32] Zhang B, Wang W and Cheng B 2015 Adv. Mech. Eng. **7** 2
[33] BagciA and Ansari R 2004 *Recognition ICPR***3** pp 2–5
[34] Luo R C, Hsu C H and Wen YC 2020 *Int. Conf. on Adv. Robo. and Intel. Syst.(ARIS)*pp 1-6
[35] Tipprasert W, Charoenpong T, Chianrabutra C andSukjamsri, C. 2019 *First International Symposium on Instrumentation, Control, Artificial Intelligence, and Robotics (ICA-SYMP)* pp 61-64

A Study on Movie Recommendations using Collaborative Filtering

Rahul Pradhan[1], **Ashish Chandra Swami**[2], **Akash Saxena**[3] **and Vikram Rajpoot**[4]

[1,4]Department of CEA, GLA University, Mathura
[2]S S Jain Subodh PG College, Jaipur
[3]Compucom Institute of Technology and Management, Jaipur

E-mail: [1]rahul.pradhan@gla.ac.in, [2]ashishchandraswami@gmail.com, [3]akash.saxena@gmail.com, [4]vikram.rajpoot@gla.ac.in

Abstract. The Recommendation system plays a major role nowadays, which is used for many applications. We know that the online content and service providers have a huge amount of content so the problem which arises is which data is required for whom so the problem of providing apposite content frequently. This paper represents the overview and approaches of techniques generated in a recommendation system. One of them is Collaborative Filtering which we will discuss about.

1. Introduction

The evolution of technology introduced the endless platform from basic to advance such as Machine Learning, Data Mining, and the Internet of Things, etc. As information technology is developing day by day, we can grab information from the internet effortlessly, but it becomes hard to use all of them appropriately. [1] To solve some of these problems big data was introduced. The big data is becoming a latest research interest in the cyber world. Big data has many characteristic which are volume, velocity, variety, value and veracity. [2] Big data is used to manage the data efficiency by pulling out some patterns and knowledge in an efficient way. But there are many users which have different taste so in this huge amount of data we have to give the user the useful content and helps them to decide the better one. To do these activities we have to rely on some automated machine so that it helps to reduce the effort of an individual. Here comes Recommendation System comes into a picture. Currently, Recommendation Systems are highly popular in our daily life like movies, news, books, shopping items, and music. Furthermore, the major companies like YouTube, Amazon, LinkedIn, Netflix use recommendation systems in social and e-commerce sites, etc. [3] It is used many technologies and algorithms to make the suggestion for an individual according to their requirement more precise and accurate. Recommendation Systems are classified in mainly three categories which are Content-based systems, the Collaborative Filtering system and the Hybrid recommendation system. Content-based systems works on the basis of the label or genre of an item. If a user watched a movie so it recommends similar movies based on director, a genre, and many more aspects. The main theory behind the collaborative filtering is that if users 'A' and 'B' have rated correspondingly in the past, then there will be an assumption that they will rate correspondingly in the future. Based on the abstract of 'closeness', 'past', and 'rating', the basic algorithm can take several forms. [4] Hybrid is the combination of Content Based and Collaborative Filtering System. However, all these systems are not a fragment of precise, and research is going on to enhance the real-time performance of these systems. [5] We are using Movie Lens data set for our research.

2. Related work

Many researchers have introduced many researches in the area of constructing recommendation systems. [6] Recommendation systems are majorly used various applications to recommend items that a customer is probably to be concerned with customer preferences. Here we are only taking collaborative filtering only.

2.1. Collaborative filtering

Collaborative Filtering is a technique which collects information in the form of rating and check similarities from the other users. For example, consider A and B are two users are watching movies online on a specific platform , both of them watched the A1 movie and then they watched the B2 movie(on different time).Now the user B has watched movie A2, so this movie will also be to A who hadn't watch that movie yet. There are mainly two types of collaborative filtering techniques Memory Based approach and Model Based approach. In Memory Based approach, the recommendation is done on the basis of past activities, whereas in the model based approach, some prediction will take place on the basis of ratings, page viewing time, clicks and so on. The following diagram represents the working of the collaborative filtering. Figure 1 shows an example of collaborative filtering. There are two approaches which are used

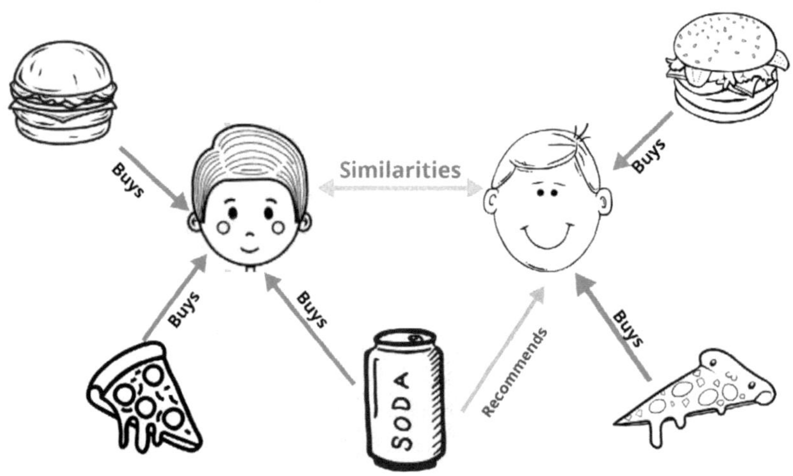

Figure 1. Collaborative Filtering Mechanism.

to solve using collaborative filtering user approach and item approach.

2.1.1. User based collaborative filtering User based collaborative filtering method is used to anticipate items to the user that are items of preference earlier for other users who are close to the user. [7] For example, let user 'A' and user 'B' have an indistinguishable preference manner. If a user A likes an item 1 user based collaborative filtering can recommend item 1 to user 2. It needs exact rating scores of items rated by users between users. Figure 2 shows a pictorial representation of the given example of a user based collaborative filtering.

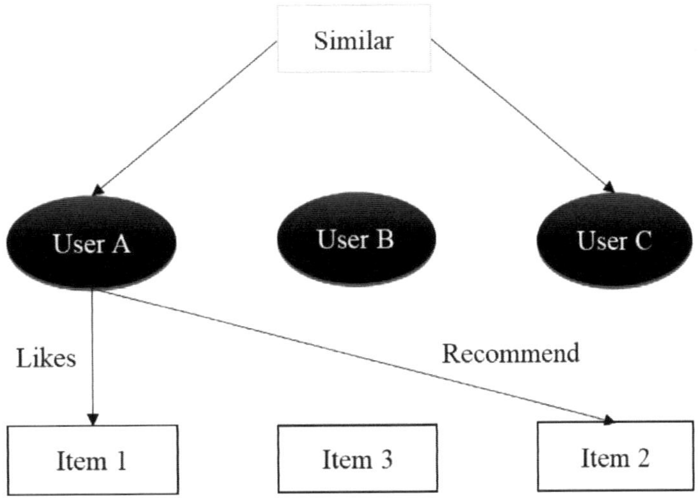

Figure 2. Explaination of user based collaborative filtering.

2.1.2. Item based collaborative filtering The item based collaborative filtering method is used to calculate items by identifying the closeness between an item and more items that are related with the user before. Let's consider item 1 and item 2 are similar. If a user likes item 1, item based collaborative filtering can suggest Item 2 to the User. It requires a kit of items that the user has rated before to anticipate the closeness between other items and a particular that item and then, it gives a suggestion in terms of that item by merging the user's earlier choices based on these item similarities [8]. In an item based collaborative filtering, a user's preferred data can be acquired in many ways like giving the rating to an item explicitly within a certain range or using various factors like click rate, users watch time, etc. Figure 3 shows a pictorial representation of the given example on item based collaborative filtering.

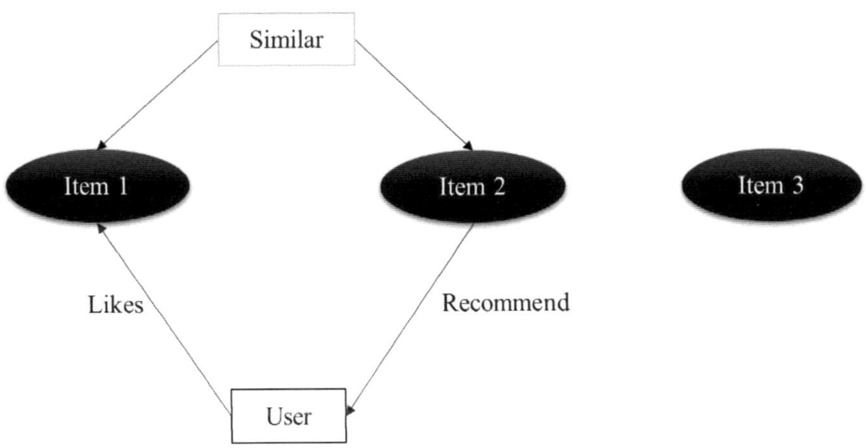

Figure 3. Explaination of item based collaborative filtering.

2.2. User similarity computation

The similarity between users is computed by estimating the value of the items estimated of the items estimated by two users. Each user uses N dimension vector to represent item score, for

IOP Conf. Series: Materials Science and Engineering 1119 (2021) 012018 doi:10.1088/1757-899X/1119/1/012018

example, to calculate of similarity of U1 and U3, first determine the set of films that they all resulted as M1,M2,M4,M5 and relative outcomes of these films. The outcome of vector U1 is 1, 3, 4, 2, and the outcome of vector u3 is 2, 4, 1, 5. The similarity of U1 and U3 is computed by the similarity formula [9]. Table 1 shows the table of above explanation.

Table 1. Calculating user's similarity.

U	M1	M2	M3	M4	M5
U1	1	3	3	4	2
U2	3	1	4		
U3	2	4		1	5
U4	2		2		

The similarity of u and u' is referred as $sim(u, u')$, the frequently used method of predicting user similarity are Cosine Similarity and Pearson Correlation similarity.

2.2.1. Cosine similarity This method computes the resemblance between two users by taking the cosine of the angle between the two vectors.

$$\begin{aligned} sim(x, y) &= cos(\overrightarrow{x}, \overrightarrow{y}) \\ &= \frac{\overrightarrow{x}.\overrightarrow{y}}{\|\overrightarrow{x}\|.\|\overrightarrow{y}\|} \\ &= \frac{\sum_{s \in s_{xy}} r_{x,s} r_{y,s}}{\sqrt{\sum_{s \in s_{x,y}} (r_{x,s})^2}.\sqrt{\sum_{s \in s_{x,y}} (r_{y,s})^2}} \end{aligned} \tag{1}$$

Among them,$r_{x,s}$ and $r_{y,s}$ are the outcome of goods s acquired by user x and y respectively, s_{xy} is the set of movies that user x and user y both acquired on. Mathematically $s_{x,y} = \{s \in items | r_{x,s} \neq \epsilon \cap r_{y,s} \neq \epsilon\}$.

2.2.2. Correlation -based similarity In this method, the resemblance between two items is computed by enumerating Pearson- r correlation. To make the correlation calculation precise we must first separate the co-rated cases. [10] Here $\overline{r_x}$ and $\overline{r_y}$, are the mean.

$$sim(x, y) = \frac{\sum_{s \in s_{xy}} (r_{x,s} - (\overline{r_x}))(r_{y,s} - (\overline{r_y}))}{\sqrt{\sum_{s \in s_{x,y}} (r_{x,s} - \overline{r_x})^2} \sqrt{\sum_{s \in s_{x,y}} (r_{y,s} - \overline{r_y})^2}} \tag{2}$$

2.2.3. Euclidean Distance Euclidean distance is the distance between two coordinates. The given equation here looks like:

$$\sqrt{(x_2 - x_1)^2 + (y_2 - y_1)^2} \tag{3}$$

Here x and y coordinates can be users and items. However, this is not a very accurate way to find the similarity between two users or items.

2.3. Pro and cons

2.3.1. Pros

- It delivers a relevant content.
- It can bring traffic to your site.
- It reduces overload.
- It reduces workload and overhead.

2.3.2. Cons (challenges faced)

- Cold start problem / Handling Unknown Users
  - When the system doesn't have much information of a user, then the system face a problem what should recommend to that user.
- Data Sparsity
  - Same as above cons. If we don't have any information about the user then how do we suggest the movies to that user?
- Scalability
  - When the number of user's data and number of items data becomes huge then we have to make sure that these data will be displayed by taking minimum time.

3. Proposed Approach

Figure 4 will tell you about the basic process involved in the recommendation system. Recommendation system works on basically on two things product details and user details. We have to collect them from the system or from the database and make decisions on the basis of ratings if a similar items were found then it will generate recommendation system otherwise no recommendation system will be generated.

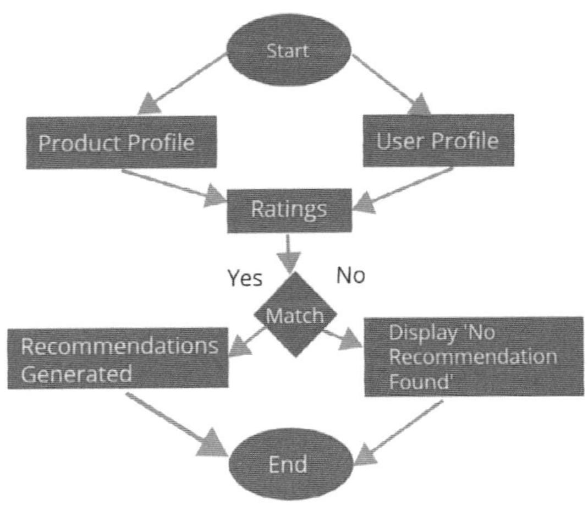

Figure 4. Flow chart for recommendation system employing collabrative filtering.

IOP Conf. Series: Materials Science and Engineering 1119 (2021) 012018 doi:10.1088/1757-899X/1119/1/012018

4. Experimental Setup

We are using Movie Lens Dataset [11] for our research. In this research we are applying item based collaborative filtering. The reason behind this is because user taste may change with respect to time but item doesn't change it remains same. There are certain stages to make our recommendation system efficiently to respond.

- Data Loading
  - To load the data and display accordingly we have to perform some operation like merging the two file in the dataset.
- Data Slicing
  - Here we are removing unnecessary column and data.
- Data Cleaning
  - In the real world data if we make a table of ratings in the recommendation system we find that most of the user are not rating the movies and are mostly inactive. The same cases are with movies either users don't watch or it's get too old. To make our computation more accurate we will remove such users and movies from our research. Now to predict similarity we have two methods either we can use correlation method or cosine method. In our research we had used correlation method.

Now to predict similarity we have two methods either we can use correlation method or cosine method. In our research we had used correlation method.

5. Results

Now let's suppose there are users who rated certain movies, for example, a certain user had rated the following movie as shown in Table 2:

Table 2. Movie Ratings.

Movie Name	User Rating
(500) Days of Summer (2009)	5
Alice in wonderland (2010)	3
Aliens (1986)	1
2001: A Space Odyssey (1968)	2

If we look carefully the user has rated romantic movies high and action movies low based this rating user will be recommended romantic movies. Here are the list of recommendation along with predictions (see Table 3). If the prediction goes to negative then that movie won't be recommended. Figure 5 shows the graph between movies and the prediction made by the recommendation system. As you can see here except few starting movies (which is rated good by user) all movies which are highly recommended from top to bottom (bottom to top on the graph 5).

IOP Conf. Series: Materials Science and Engineering 1119 (2021) 012018 doi:10.1088/1757-899X/1119/1/012018

Table 3. Predicted Movie Ratings

Movie Name	Predicted User Rating
(500) Days of Summer (2009)	2.584556
Alice in Wonderland (2010)	1.395229
Silver Linings Playbook (2012)	1.254800
Yes Man (2008)	1.116264
Adventure land (2009)	1.112235
Marley & Me (2008)	1.108381
About Time (2013)	1.102192
Crazy, Stupid, Love. (2011)	1.088757
50/50 (2011)	1.086517
Help, The (2011)	1.075963
Up in the Air (2009)	1.053037
Holiday, The (2006)	1.034470
Friends with Benefits (2011)	1.030875
Notebook, The (2004)	1.025880
Easy A (2010)	1.015771
Secret Life of Walter Mitty, The (2013)	0.997979
Perks of Being a Wallflower, The (2012)	0.967425
Toy Story 3 (2010)	0.963276
Ugly Truth, The (2009)	0.959079
Harry Potter and the Half-Blood Prince (2009)	0.954180

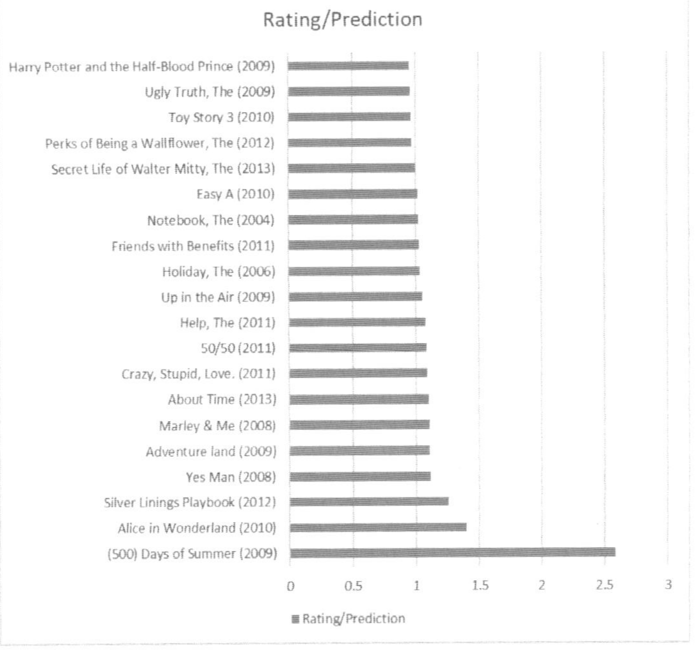

Figure 5. Graph between movies and the recommendation.

IOP Conf. Series: Materials Science and Engineering 1119 (2021) 012018 doi:10.1088/1757-899X/1119/1/012018

6. Conclusion

The Recommendation system is a very powerful technology which helps people to find what they like. These systems have certain limitations not recommending efficiently to the users. If we take Collaborative Filtering as an example, then we find it is most successful and powerful algorithm, [12] even though it is best but this algorithm has some high runtime and faces some issues like data Sparsity which can be removed by using a Hybrid movie recommendation system. But each algorithms have its strength and weakness. Our proposed approach can handle quite a big amount of data effectively. In future, we will work on its weakness and on its user interface.

References

[1] Ko S K, Choi S M, Eom H S, Cha J W, Cho H, Kim L and Han Y S 2011 *Symposium on Human Interface* (Springer) pp 558–566
[2] Aljunid M F and Manjaiah D 2019 *Data Management, Analytics and Innovation* (Springer) pp 283–295
[3] Wang Z, Yu X, Feng N and Wang Z 2014 *Journal of Visual Languages & Computing* **25** 667–675
[4] Goel D and Batra D 2009 *Department of Electrical and Computer Engineering, Carniege Mellon University* 1–7
[5] Raval N and Khedkar V 2019 *INTERNATIONAL JOURNAL OF SCIENTIFIC & TECHNOLOGY RESEARCH* **8** 2507–2512
[6] Wei J, He J, Chen K, Zhou Y and Tang Z 2016 *2016 IEEE 14th Intl Conf on Dependable, Autonomic and Secure Computing, 14th Intl Conf on Pervasive Intelligence and Computing, 2nd Intl Conf on Big Data Intelligence and Computing and Cyber Science and Technology Congress (DASC/PiCom/DataCom/CyberSciTech)* (IEEE) pp 874–877
[7] Lee Y 2015
[8] Gong S 2010 *JSW* **5** 745–752
[9] Cui B B 2017 *ITM web of conferences* vol 12 (EDP Sciences) p 04008
[10] Sarwar B, Karypis G, Konstan J and Riedl J 2001 *Proceedings of the 10th international conference on World Wide Web* pp 285–295
[11] MovieLens 2018 Introduction-to-Machine-Learning https://github.com/codeheroku/Introduction-to-Machine-Learning/tree/master/CollaborativeFiltering/dataset (Accessed on 01/31/2021)
[12] Phorasim P and Yu L 2017 *International Journal of Advanced Computer Research* **7** 52

Image segmentation with Kapur, Otsu and minimum cross entropy based multilevel thresholding aided with cuckoo search algorithm

R Kalyani[1], P D Sathya[2] and V P Sakthivel[3]

[1,2] Department of Electronics and Communication Engineering, Faculty of Engineering & Technology, Annamalai University, Tamilnadu, India – 608002.
harebalu@gmail.com
[3]Department of Electrical and Electronics Engineering, Government College of Engineering, Dharmapuri, Tamilnadu, India
[1]vp.sakthivel@yahoo.com

Abstract. Color image segmentation is the primary factor to provide the intended information from the input image. The straightforward method called multilevel thresholding (MLT) is used to analyse the various classes of complex images. But, when the level of threshold increases, computational difficulty increases. Hence, MLT with most promising objective functions such as Kapur, Otsu and minimum cross entropy aided with cuckoo search algorithm (CSA) is used. The efficient metaheuristic cuckoo search algorithm's controlling parameter balances the local and global search. In this paper, the efficacy of CSA at 4,5,6 and 7 threshold levels with various fitness functions are utilized for precise image segmentation. It is seen from experimental results, the Otsu based cuckoo search algorithm outperform than Kapur and MCE. Quality metrices such as computational time, PSNR (peak signal to noise ratio) and SSIM (structural similarity index) authenticate the exploration and exploitation capability of CSA algorithm for real-world applications.

1. Introduction

Color image segmentation is the basic pre-processing step to examine the image and it finds applications in video processing, medical analysis, computer vision, industrial production and so forth [1]. Various segmentation techniques namely thresholding, edge detection, region based are most widely used as an initial step in image processing [2]. Among these techniques, thresholding is the straight- forward technique to retrieve the information we require. Bilevel thresholding classifies the simple image into object of interest and the background with single threshold value, whereas various classes of complex images are obtained through multilevel thresholding with various threshold levels [3]. The maximising objective functions such as Otsu method calculates the optimal threshold by maximising between the class variance, Kapur entropy gains maximum information by estimating the pixel relationship and minimum deviation between the true and predicted image is obtained by minimum cross entropy method [4-7].

Optimisation plays a vital role in finding feasible solutions for various engineering problems. Most of the bio-inspired algorithms use tuning parameters to locate the global threshold for perfect image segmentation. Generally, conventional methods get trapped in sub-optimal solutions without reaching the target point. Thus, to overcome such complexities, metaheuristic algorithms such as differential evolution (DE) helps the fitness function to reach the goal by controlling the parameters of

DE [8]. Ant colony optimisation (ACO) provides perfect segmentation using entropy method. Inspired by Darwin evolution, Genetic algorithm (GA) finds wide applications in optimisation problems, but GA fails in controlling the parameters and get stuck with local point [9], Particle swarm optimisation (PSO) mimics the behaviour of birds and fishes and it is easy to implement but results in limited search space due to fixed trajectory [10]. Honey-bee algorithm with Otsu, Bayesian produce improved image segmentation [11]. Precise classification is attained by improved variational mode decomposition for fast target recognition compared to PSO, BF (Bacterial foraging) [12] etc. In the listed aforesaid bio-inspired algorithms require controlling of various parameters. Hence, selection of suitable parameters to seek optimal values is essential. The most widely used efficient, cuckoo search algorithm falls under the bio-inspired algorithm category to increase their population through reproductive strategy and it is computationally effective than PSO [13]. Cuckoo search algorithm mimics the cuckoo birds and their strategy, laying eggs in another bird's nest. The host bird hurls the eggs away if the eggs are alien or it builds a new nest leaving the old nest. CSA assumes that every cuckoo lay eggs one by one. The best eggs are chosen for the next round fixing the number of nests. According to the literature the cuckoo search algorithm is better than other methods. CSA employs Levy fight strategy, has excellent global and local search capabilities and hence CSA is used in wide range of applications.

2. Problem estimation of multilevel thresholding methods

Kapur entropy method, Otsu (between-class variance method) and minimum cross entropy method [14-15] for accurate image segmentation are widespread in use. Thresholding with more than single thresholds are called multi-level thresholding and represented as:

$$o_1(x,y)=\{i(x,y)\in I\,|\,0\le I(x,y)\le t_1-1\} \quad (1)$$

$$o_2(x,y)=\{i(x,y)\in I\,|\,t_1\le I(x,y)\le t_2-1\} \quad (2)$$

$$o_i(x,y)=\{i(x,y)\in I\,|\,t_i\le I(x,y)\le t_{i+1}-1\}\ldots \quad (3)$$

$$o_r(x,y)=\{i(x,y)\in I\,|\,t_r\le I(x,y)\le L-1\} \quad (4)$$

where $t_1,t_2,t_3,t_4\ldots t_i,\ldots t_r$ indicate various thresholds, *I(x,y)* stands for input image and *o(x,y)* indicate output image. In this proposed paper, the part to be explored from the color image, is interpreted by using maximising objective functions such as Otsu, Kapur and with minimum entropy function MCE.

2.1 Kapur method

Kapur entropy is most widely used fitness function for color image segmentation through multilevel thresholding method [16].

Maximizing the objective function by:

$$f(t)=F_0+F_1 \quad (5)$$

$$F_0=-\sum_{i=0}^{t-1}\frac{P_i}{X_0}\ln\frac{P_i}{X_0};\quad X_0=\sum_{i=0}^{t-1}P_i \quad (6)$$

$$F_1=-\sum_{i=t}^{G-1}\frac{P_i}{X_1}\ln\frac{P_i}{X_1};\quad X_1=\sum_{i=t}^{G-1}P_i \quad (7)$$

Thus, Kapur's entropy achieves unification of the histogram for image segmentation. Extension of Kapur's concept for Multilevel thresholding:

For 'k' dimensional optimization problem, 'k' optimal thresholds of an image $[t_1,t_2,\ldots t_k]$ to maximise the objective function.

$$f[t_1,t_2,\ldots t_k]=F_0+F_1+\cdots+F_k \quad (8)$$

where

$$F_0 = -\sum_{i=0}^{t_1-1} \frac{P_i}{X_0} \ln \frac{P_i}{X_0}; \quad X_0 = \sum_{i=0}^{t_1-1} P_i \tag{9}$$

$$F_1 = -\sum_{i=t_1}^{t_2-1} \frac{P_i}{X_1} \ln \frac{P_i}{X_1}; \quad X_1 = \sum_{i=t_1}^{t_2-1} P_i \tag{10}$$

$$F_2 = -\sum_{i=t_2}^{t_3-1} \frac{P_i}{X_2} \ln \frac{P_i}{X_2}; \quad X_2 = \sum_{i=t_2}^{t_3-1} P_i ,\dots \tag{11}$$

$$F_k = -\sum_{i=t_k}^{G-1} \frac{P_i}{X_k} \ln \frac{P_i}{X_k}; \quad X_k = \sum_{i=t_k}^{G-1} P_i \tag{12}$$

2.2 Otsu's method

The between- class variance (Otsu's) criteria predict the optimal threshold by maximising between-class variance [17].

According to Otsu's between-class variance discriminant analysis:

$$y(t) = \sigma_0 + \sigma_1 \tag{13}$$

$$\sigma_0 = x_0(\mu_0 - \mu_T)^2 \tag{14}$$

$$\sigma_1 = x_1(\mu_1 - \mu_T)^2 \tag{15}$$

Otsu's bilevel optimal threshold 't*' as

$$t* = arg\,max\{y(t)\} \qquad 0 \le t \le G-1 \tag{16}$$

Extension of Otsu's concept for multilevel thresholding is represented as:

$$y(t) = \sigma_0 + \sigma_1 + \sigma_2 + \cdots + \sigma_k \tag{17}$$

$$\sigma_0 = x_0(\mu_0 - \mu_T)^2, \tag{18}$$

$$\sigma_1 = x_1(\mu_1 - \mu_T)^2,\dots \tag{19}$$

$$\sigma_k = x_k(\mu_k - \mu_T)^2 \tag{20}$$

2.3 Minimum cross entropy

Minimum discrimination information is the summation of entropy and its divergence [17].

Cross entropies minimum objective function is stated as:

$$\min \{D(m)\} = D_0 + D_1 \tag{21}$$

$$D_0 = -\sum_{i=0}^{m-1} ih(i) \log\left(\frac{\sum_{i=0}^{m-1} ih(i)}{\sum_{i=0}^{m-1} h(i)}\right) \tag{22}$$

$$D_1 = -\sum_{i=m}^{G} ih(i) \log\left(\frac{\sum_{i=0}^{G} ih(i)}{\sum_{i=0}^{G} h(i)}\right) \tag{23}$$

Complex image information is received by multilevel thresholding and for determining 'm' dimensional optimization, the objective function is considered as:

$$\min D(m_0 + m_0 + m_0 + ... + m_0) = D_0 + D_1 + D_2 + ... + D_n \quad (24)$$

where,

$$D_0 = -\sum_{i=0}^{m_1-1} ih(i)\log\left(\frac{\sum_{i=0}^{m_1-1} ih(i)}{\sum_{i=0}^{m-1} h(i)}\right) \quad (25)$$

$$D_1 = -\sum_{i=m_1}^{m_2-1} ih(i)\log\left(\frac{\sum_{i=m_1}^{m_2-1} ih(i)}{\sum_{i=m_1}^{m_2-1} h(i)}\right) \quad (26)$$

$$D_2 = -\sum_{i=m_2}^{m_3-1} ih(i)\log\left(\frac{\sum_{i=m_2}^{m_3-1} ih(i)}{\sum_{i=m_2}^{m_3-1} h(i)}\right) .... \quad (27)$$

and

$$D_n = -\sum_{i=n}^{G} ih(i)\log\left(\frac{\sum_{i=n}^{G} ih(i)}{\sum_{i=n}^{G} h(i)}\right) \quad (28)$$

With the increase in number of thresholds, computational time goes up, limiting the multilevel thresholding applications. The above problem is overcome by predicting the perfect parameters of Otsu, Kapur and MCE multilevel thresholding using EMA algorithm for excellent medical image segmentation. The proposed method maximizes the Kapur and Otsu's fitness function, while minimizing the MCE function.

3. Cuckoo Search Algorithm

Cuckoo birds show some collective emerging and self-organising characteristics. This algorithm solves NP-hard problems. The metaheuristic cuckoo search algorithm was conceived by Yan and Deb in 2009 [18]. Cuckoo search algorithm is used to find an optimised solution with cuckoo eggs. The cuckoo search birds are fascinating, and they lay 16 to 22 eggs in communal nests. Timing of laying the eggs by sporting the nests exactly shows the unique reproductive approach of cuckoo birds. The cuckoo bird lay only one egg in their host nest by quick hatching. Rest of the unhatched eggs will be removed by foreign cuckoo. The time cuckoos find that the eggs do not belong to them, will be thrown out and abandoned to build a new nest. This strategy improves their hatching probability. This method uses levy flight to generate random steps and choose random direction with step length.

Thus, the algorithm depends on cuckoo's parasitic reproduction strategy and Levy flight global and local search principle.

3.1 Levy flights

The random walks performed by animals and insects in random direction is called Levy flights and the Levy direction is driven by the step length. Sudden 90-degree turns are associated with the walks and Levy flight variance increases faster and thereby decreases the algorithmic iterations than other random walks.

Intensified search by exploiting the limited search area is achieved through local random walk and this is expressed as

$$x_p^{d+1} = x_p^d + \alpha s \otimes H(P_a - \varepsilon) \otimes (x_q^d - x_r^d) \quad (29)$$

$$x_p^{d+1} = x_p^d + \alpha L(s, \lambda) \quad (30)$$

Here, x_p^d and x_r^d indicates current positions by random permutation, α is the positive step size factor, x_p^{d+1} indicates the updated position, s denotes the step size, $\otimes$ represents the product of two vectors, H

is the Heavy-side function, $P_a \in [0,1]$ and P_a the switching parameter controls the search between local and global threshold, Ɛ is the random number and $L(s,\lambda)$ is the Levy distribution to determine the step size.
Each egg in cuckoo search algorithm represents a solution and thus each cuckoo can lay only one egg indicating only one solution. Step length is much longer in the longer run and thus Levy flight explores the search space efficiently and effectively. Levy flight, randomized step length and less tuning parameters are the main advantage of cuckoo search algorithm for wide range of applications such as nurse scheduling problems by Lim Huai Tein [19], Layeb [20] quantum computing and principles, Traveling salesman problem by Aziz [21] and so forth.

4. Algorithmic steps to implement CSA for MLT using Kapur, Otsu and MCE objective function

Search step 1: *Initialization*

Initialise randomly the M bird's nest location as $X = (x_1, x_2, \ldots, x_M)$ and the M bird's nest positions are fed to test through objective functions in Eq. 8,13 and 21. The output of the functions decides the nest location to carry it over the next generation.

Search step 2: Exploitation of the search space

The local random walks are updated by the Eq.29. By refining in limited search space, improvement in current solution is obtained.

Search step 3: Exploring the location

The global random walks are revised by the Eq.30. Wide search is carried out by exploring the promising solutions. Thus, diversified search avoids getting trapped with local point.

Search step 4: Update the position

Evaluate the fitness function's current output with exploration capability and update the new solution until the maximum number of iterations is reached.

Search step 5: Optimal threshold output is obtained by reaching the best nest and its flow chart is represented in figure 1.

Cuckoo search pseudocode:

begin
Initialise the variables such as population size (nest: x_i = 1, 2, ..., M)
while stopping condition is not arrived
Randomly develop a cuckoo by means of levy flight
Evaluate the fitness function F_c according to Eq.8,13 and 21.
Choose the random nest within x_i and evaluate its suitability as (F_j)
if $F_c > F_j$
update the new solution according to if condition
end if
Drop out the fraction of alien nest p_a
Aid levy flight to locate new nest at new spot
Retain the leading nests
Rank and compute the current best nest
end while
Get the computed output
end

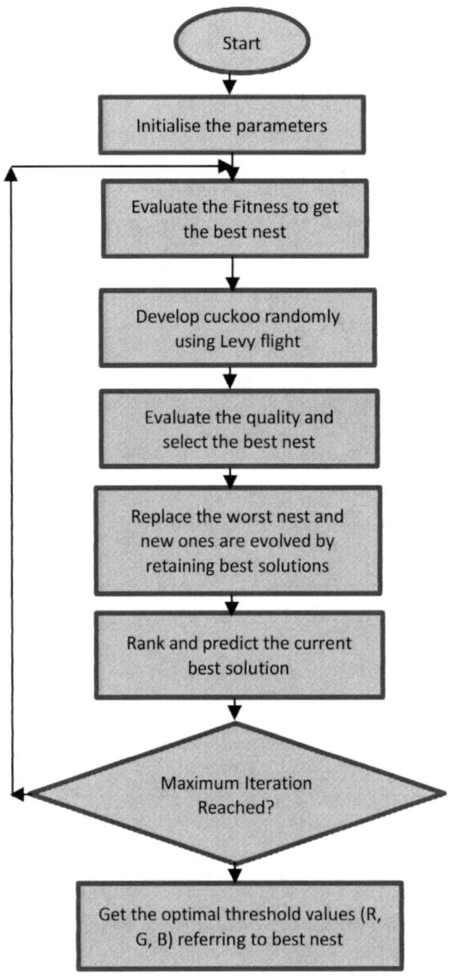

Figure 1. Flow chart of Image segmentation of CSA algorithm

5. Experimental Results

Cuckoo search algorithm based multilevel thresholding simulations have been examined in MATLAB 7.0, processed in Intel core 2 Duo Processor(3GHz), 2 GB RAM. Performance of CSA based MLT aided with Kapur, Otsu and MCE objective functions are used to analyze the segmentation quality. Color test input images in Fig. 2. such as Lena (512×512), Airplane (512×512), Baboon (512×512), Goldhill (720×576) and Starfish (480×512) along with their histogram are illustrated and these images are taken from Berkeley segmentation and COCO dataset. Cuckoo search algorithm is computed at 4-level, 5-level, 6-level, and 7-level thresholds. Optimization is authenticated by best fitness value CPU time and convergence rate. Furthermore, segmented image quality is predicted by peak signal to noise ratio (PSNR) and structural similarity index (SSIM).

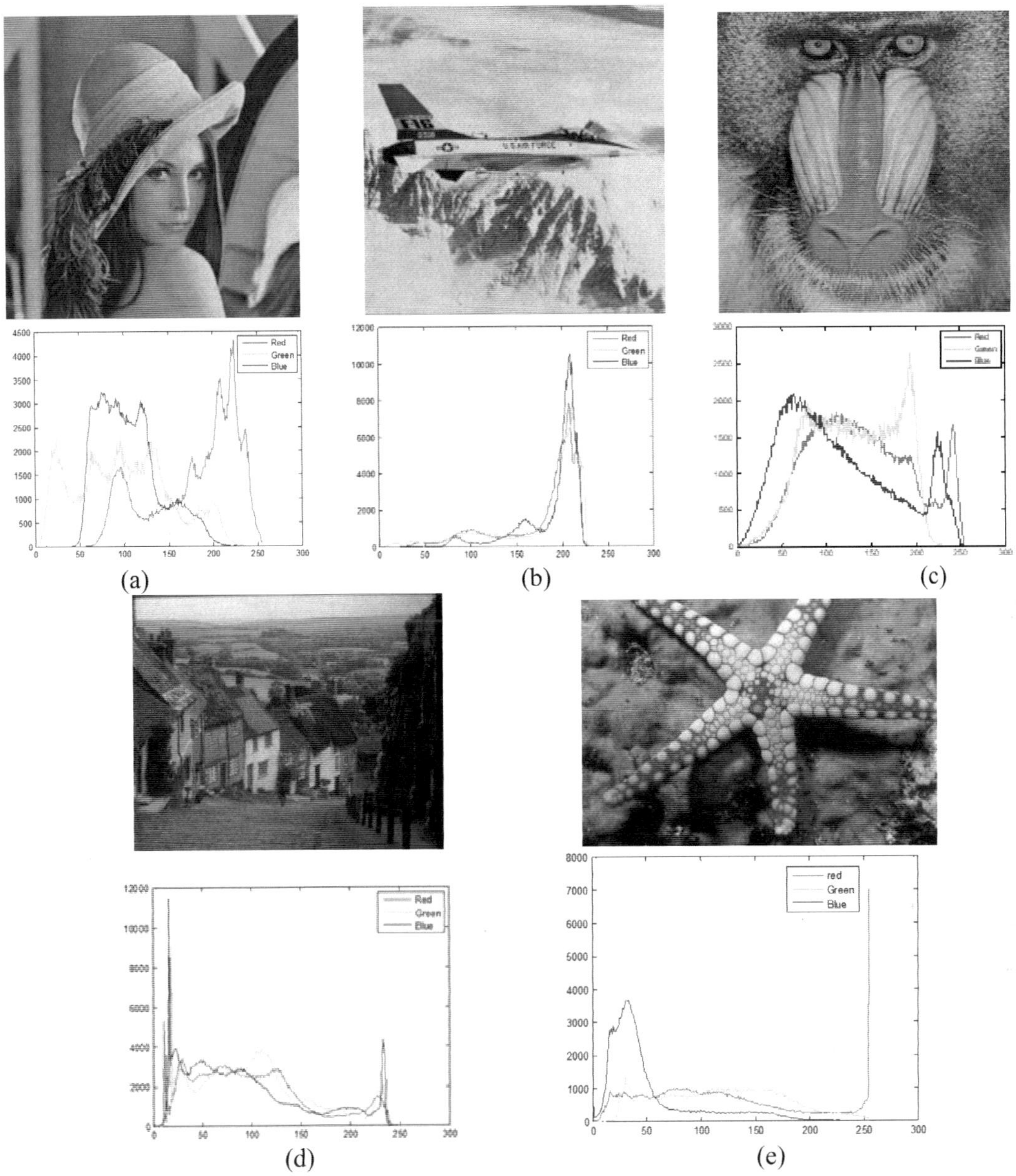

Figure 2. Standard test color images and their histograms

(a)Lena (b) Airplane (c) Baboon (d) Goldhill (e) Starfish

5.1 Performance evaluation by Kapur's method

The segmented output of cuckoo search algorithm at 4,5,6,7 threshold level for various input images is in figures 3. Table 1 shows that CSA algorithm based on Kapur method. Comparing Lena, Airplane, Baboon, Goldhill and Starfish images, the output segmented image of Baboon performed well for all the threshold levels of CSA. This makes the CSA based color segmentation easy even with higher threshold levels. Airplane image segmentation took less computational time for 4th CSA threshold level, similarly, Lena image for 5th and 7th CSA threshold level, Starfish for 6th CSA threshold level. The best objective values of CSA reduce the time to explore and exploit the threshold value. The switching parameter Pa set as 0.25 with step size 0.01 balances the best output at lower and higher threshold levels. Thus, the switching parameter variable switches between the exploration and exploitation to achieve superior output. The accurate overall outperformance of CSA overcomes the drawback of getting stuck with sub-optimal solutions.

Table 5 shows the performance of CSA in terms of PSNR to specify the accuracy of an image. The region of interest is inferred by avoiding over and under segmentation through the best PSNR output of CSA based MLT. PSNR table shows that accurate image segmentation of Lena at 4 and 6 threshold levels, Starfish at 5th and Baboon at 7th threshold level. Thus, High PSNR with low MSE affirms the low degree of distortion in the image. True and segmented image consistency is obtained through SSIM (Structural similarity index). Table 1 shows the better performance of Lena at 4,5 and 7th level and Baboon at 6th threshold level.

Table 1. Optimal Threshold and Objective values of each algorithm under Kapur method

Input Image	No. of thresholds	Red Band	Green Band	Blue Band	Objective values
Lena	4	103 142 172 193	85 103 126 162	95 108 144 198	51.779538
	5	112 132 192 215 229	57 81 99 193 207	82 105 139 151 167	57.966487
	6	102 122 146 183 188 233	35 92 150 167 190 226	91 132 137 169 201 213	63.906785
	7	67 102 115 141 194 203 226	48 109 141 152 161 190 195	78 107 127 153 159 184 210	70.274729
Airplane	4	65 86 151 209	84 119 167 194	65 138 171 208	51.924100
	5	52 111 127 184 208	85 111 158 202 210	63 121 132 152 200	58.340205
	6	68 123 142 177 187 201	49 121 137 155 187 203	62 118 165 183 206 223	64.867523
	7	48 72 112 133 159 184 205	43 96 145 177 190 196 217	32 57 117 149 166 179 216	71.707438
Baboon	4	47 125 178 188	30 68 103 151	43 87 170 210	55.185184
	5	38 101 160 188 226	45 109 129 159 175	78 109 138 211 227	62.811760
	6	67 104 129 147 214 242	55 108 127 151 181 206	57 123 138 189 204 228	69.402750
	7	49 108 128 166 177 200 216	30 78 104 136 162 190 214	55 64 77 117 178 205 221	77.053829
Goldhill	4	61 88 156 185	99 127 195 218	53 81 132 203	54.905472
	5	77 135 168 204 220	47 80 129 163 192	113 127 169 185 197	61.862014
	6	53 73 132 158 176 193	54 79 92 141 195 210	61 83 108 134 173 179	69.180404
	7	56 71 88 106 135 166 199	59 80 148 162 185 212 230	41 56 81 102 143 167 224	76.121606
Starfish	4	79 133 175 225	75 104 179 226	52 139 182 209	55.123595
	5	54 90 109 165 226	106 149 173 218 238	47 121 138 181 199	62.008888
	6	41 86 100 175 207 245	101 134 158 191 218 241	63 83 134 174 190 209	69.440745
	7	48 71 100 145 174 202 237	78 114 140 158 166 216 238	68 88 133 156 175 208 213	76.285343

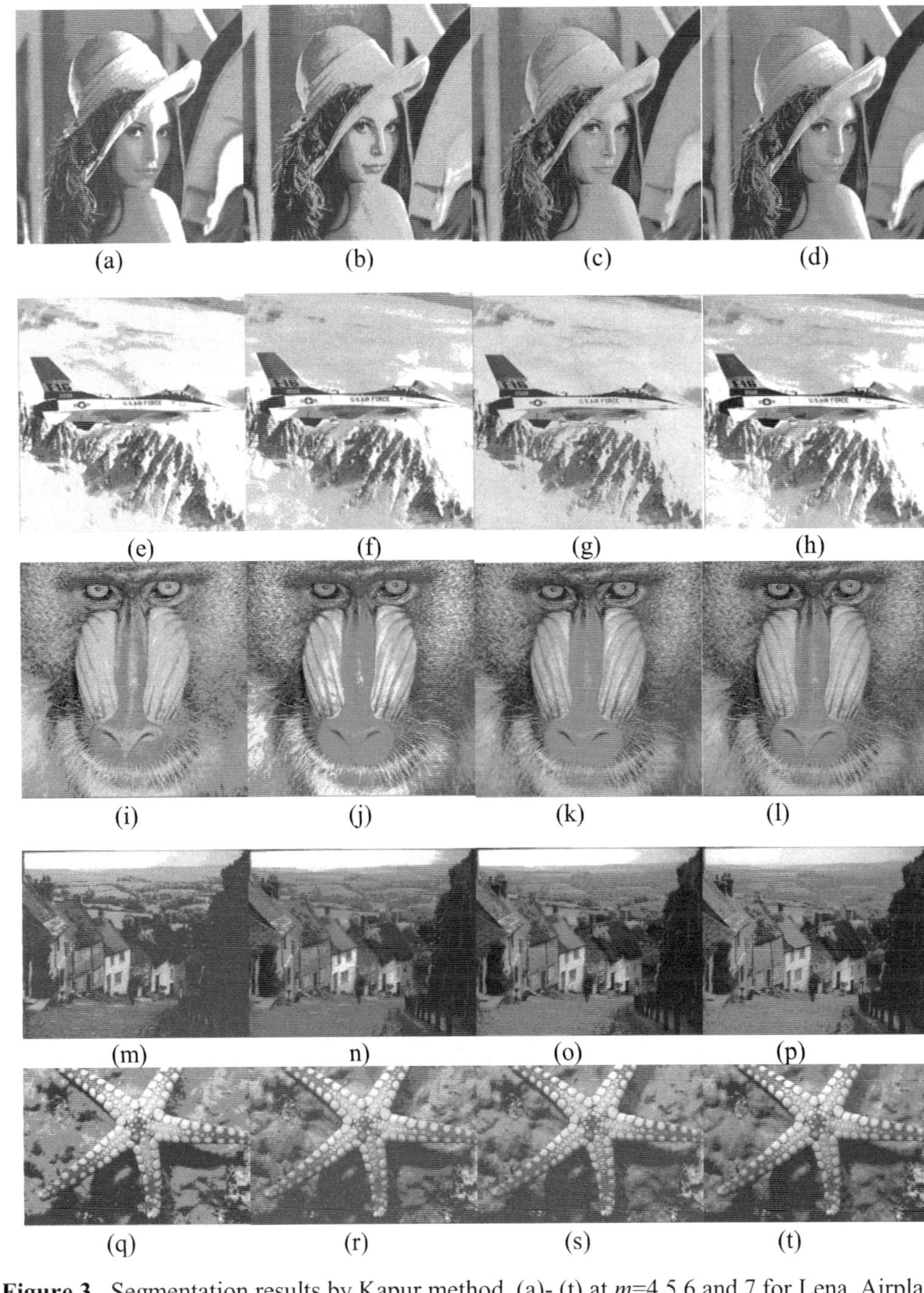

Figure 3. Segmentation results by Kapur method. (a)- (t) at m=4,5,6 and 7 for Lena, Airplane, Baboon, Goldhill and Starfish images

5.2 Performance evaluation based on Otsu's method

Figure 4 shows the segmented output of different algorithms using Otsu method at 4,5,6 and 7th threshold level. Table 2 lists the fitness function values and selected threshold values based on Otsu's entropy. It is very clear that the Otsu method's result is consistent compared to Kapur's entropy. For example, the baboon objective values prove the accurate image segmentation of Airplane image in Otsu's method with the values 16678.75537, 16695.57938, 16711.16156 and 16730.00905 at 4,5,6 and

7 threshold levels, respectively. The reason behind CSA achieving the best segmented output is mainly using Levy flight strategy to search for unknown area quickly without any deviation. The Otsu based CSA wins with excellent results compared to Kapur and MCE indicating the excellent intensification and diversification capability. Table 2 of Otsu based EMA, indicates the Baboon, Airplane, Lena, and Starfish images at 4,5,6 and 7 threshold levels respectively, winning with least execution time through tactical control of market risk variables g1 and g2. Detailed information through PSNR and SSIM from the Table 4 and Table 6 affirm the best performance of Otsu based CSA.

Table 2. Optimal Threshold and Objective values of each algorithm under Otsu method

Input Image	No. of thresholds	Red Band	Green Band	Blue Band	Objective values
Lena	4	111 132 167 222	37 99 139 176	52 96 127 154	10000.93306
	5	114 118 154 193 244	50 74 106 118 159	90 139 167 185 231	10021.79832
	6	108 131 143 195 217 232	44 70 83 128 163 192	52 91 132 140 178 199	10050.55737
	7	38 112 138 166 197 214 228	49 104 138 144 159 210 218	75 94 120 144 171 213 239	10063.47963
Airplane	4	83 148 188 210	43 94 153 183	35 112 177 203	16678.75537
	5	84 139 175 183 206	51 87 103 142 201	110 116 177 205 251	16695.57938
	6	92 141 146 151 195 234	48 90 140 179 199 218	117 146 164 171 199 203	16711.16156
	7	78 109 130 159 170 198 208	77 91 127 159 187 204 237	35 114 137 177 193 201 221	16730.00905
Baboon	4	73 122 172 234	92 102 150 187	77 136 160 194	10120.50075
	5	83 136 173 183 273	71 82 123 165 218	52 99 141 150 205	10159.04394
	6	97 135 168 188 200 221	83 113 120 130 152 169	69 103 127 153 172 229	10186.45124
	7	68 118 123 157 196 212 218	63 114 136 148 164 183 197	47 79 97 133 166 192 217	10218.74905
Goldhill	4	57 101 157 163	62 104 166 219	51 110 183 222	7473.054615
	5	82 112 178 202 227	45 60 111 130 194	62 93 122 165 204	7500.457409
	6	39 67 108 143 165 191	57 95 106 136 153 194	72 91 129 156 175 199	7541.36196
	7	41 66 93 99 122 165 170	51 95 131 168 179 186 208	44 48 62 90 98 135 199	7582.882381
Starfish	4	53 113 142 203	58 130 161 215	43 54 81 156	7450.00132
	5	54 103 159 192 242	47 131 158 195 238	38 84 140 178 232	7483.21717
	6	37 93 108 146 189 225	36 58 118 122 183 242	37 47 51 99 135 174	7497.254725
	7	52 64 93 145 161 186 230	61 102 128 157 184 218 230	53 95 123 147 181 197 215	7540.674017

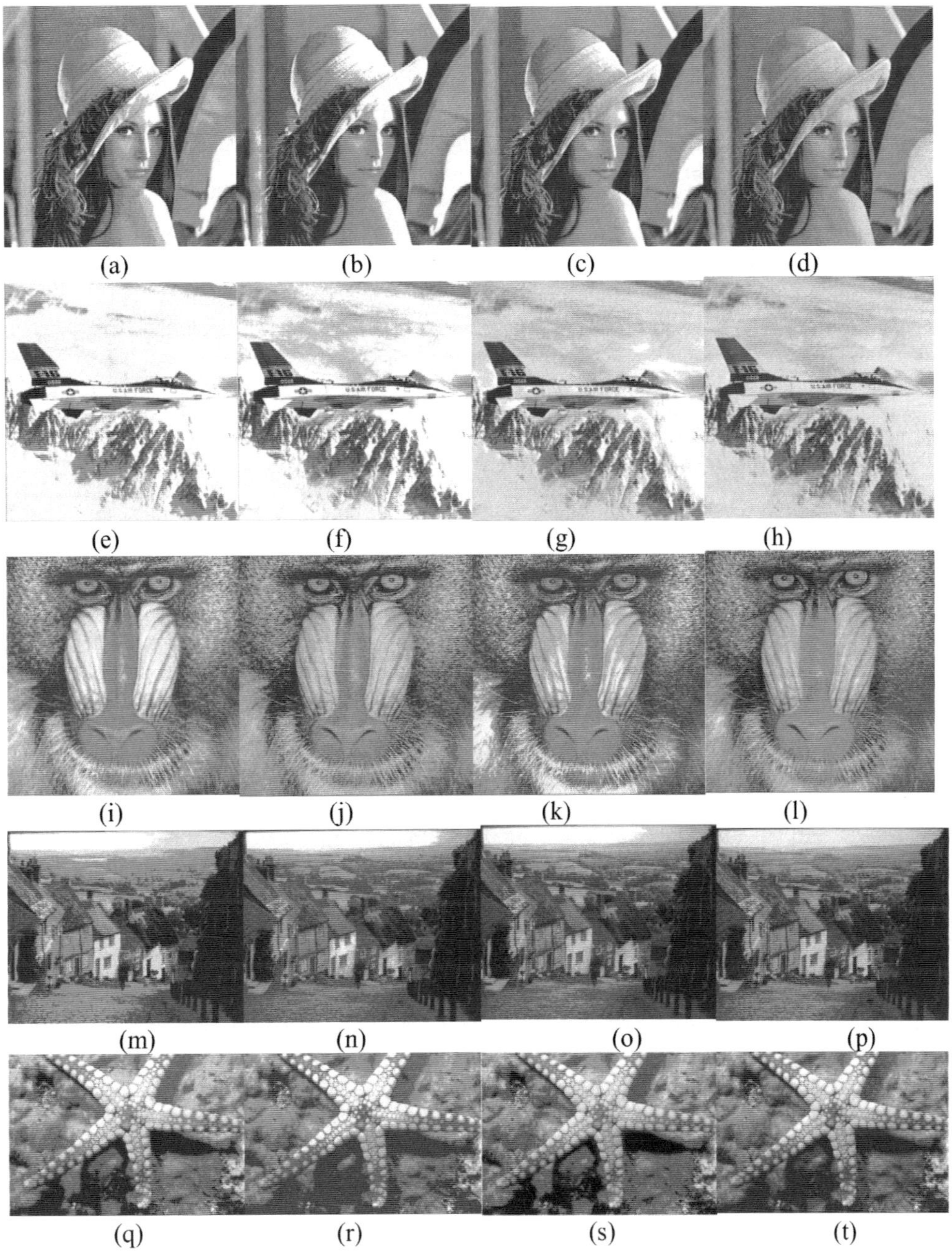

Figure 4. Segmentation results by Otsu method. (a)- (t) at m=4,5,6 and 7 for Lena, Airplane, Baboon, Goldhill and Starfish images

5.3 Performance evaluation based on MCE method

The segmented performance of CSA in terms of optimal threshold values, objective function values in Table 3 and its segmentation results are in figure 5. Due to stochasticity in metaheuristic algorithms, the CSA experiments are run for 100 times. It can be confirmed from the table that the better segmentation is obtained for Airplane at 4th and 7th level and Baboon images at 5th and 6th threshold level. Optimal threshold selection process is generally considered as constrained optimization problem. Thus, the image segmentation quality is determined by CSA based metaheuristic algorithms. In this paper, Table 5 depicts that the CSA algorithm achieves the least CPU time Starfish (4th and 7th threshold level) and Lena at 5th and 6th compared to other considered images. Table 4 and Table 6 show the Higher PSNR and SSIM of CSA based MCE predicting the segmentation quality.

Table 3. Optimal Threshold and Objective values of each algorithm under MCE method

Input Image	No. of thresholds	Red Band	Green Band	Blue Band	Objective values
Lena	4	99 120 182 213	88 116 167 233	73 121 144 183	28703.09866
	5	114 141 185 208 223	103 128 148 191 251	87 135 170 197 224	24577.15000
	6	97 111 176 189 204 229	30 82 136 172 213 243	83 99 119 137 178 236	21612.70379
	7	74 90 141 167 193 217 235	60 84 120 121 155 205 224	80 110 119 129 171 216 244	19118.64986
Airplane	4	82 149 176 205	118 150 195 240	90 134 163 201	33939.77180
	5	80 149 185 196 210	103 159 196 207 233	109 131 197 205 229	27098.50875
	6	115 136 143 169 197 221	76 107 148 197 212 220	99 146 160 185 207 221	22452.43679
	7	70 117 147 176 185 202 216	79 81 140 165 184 195 213	53 111 138 158 199 204 240	21538.68202
Baboon	4	117 149 190 214	107 150 178 196	51 105 157 184	31227.48433
	5	69 120 177 202	93 133 160 188 202	97 149 165 182 239	27468.89777
	6	101 127 156 173 214 234	80 126 172 180 193 241	61 100 155 189 213 221	24029.79900
	7	71 102 128 148 189 207 240	38 84 104 146 164 188 225	40 95 120 150 164 195 221	20110.63566
Goldhill	4	80 108 129 172	80 113 135 169	98 142 151 204	24318.98900
	5	69 81 112 152 203	76 109 122 180 217	104 119 186 207 238	21969.71380
	6	51 91 105 131 191 217	61 123 153 179 219 228	57 111 170 198 218 225	19111.57907
	7	57 89 109 143 201 228	77 116 137 149 158 172 193	77 110 142 177 195 213 222	16099.62704
Starfish	4	92 127 171 221	94 147 172 208	98 100 161 210	23545.85164
	5	76 96 133 190 238	110 138 185 203	69 128 158 176 199	20656.05822
	6	97 129 149 176 207 241	95 128 137 182 206 213	47 103 139 143 191 240	18332.34007
	7	78 105 124 153 166 204 245	74 112 149 178 185 203 230	45 88 117 142 180 207 251	14642.14150

IOP Conf. Series: Materials Science and Engineering 1119 (2021) 012019 doi:10.1088/1757-899X/1119/1/012019

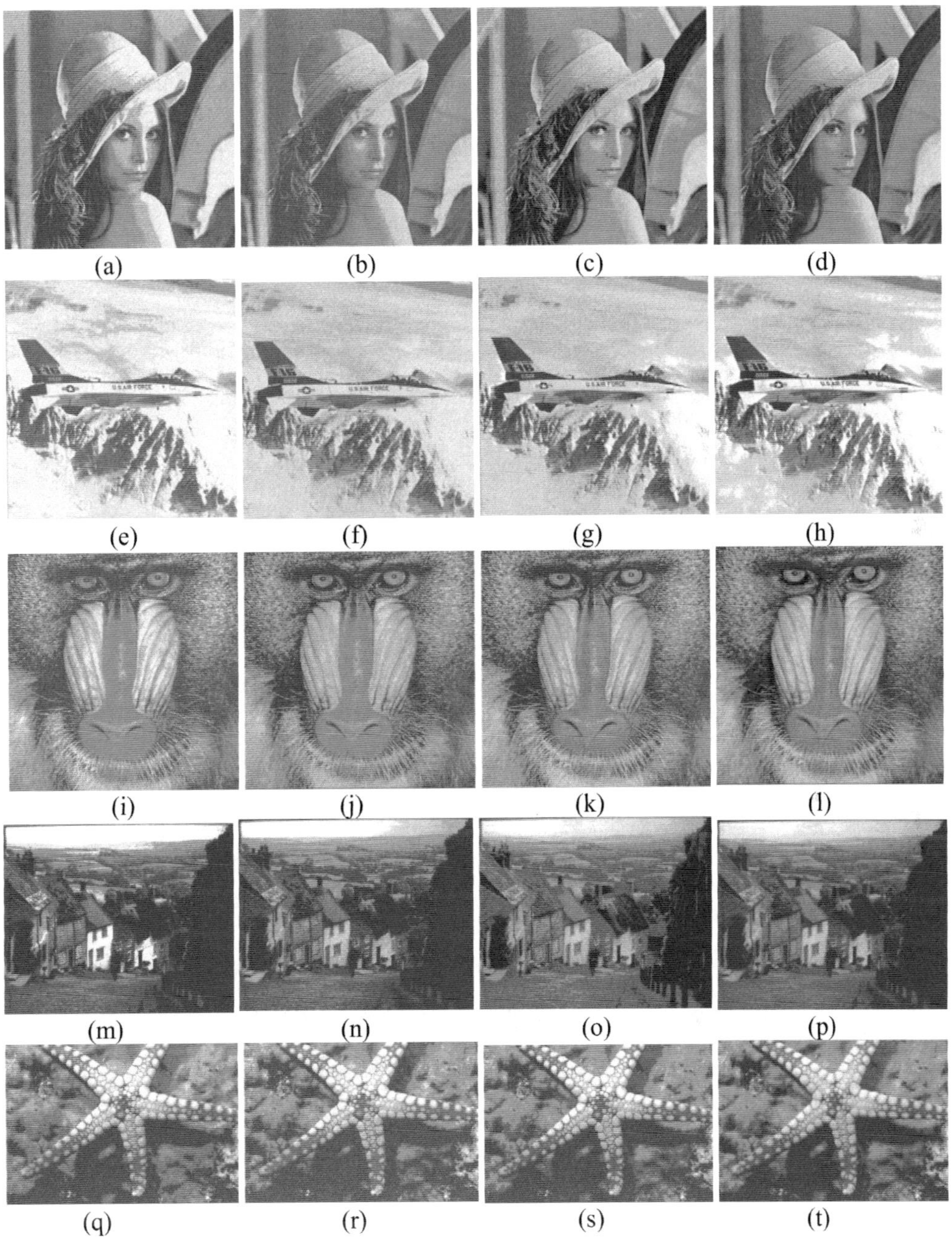

Figure 5. Segmentation results by MCE method. (a)- (t) at m=4,5,6 and 7 for Lena, Airplane, Baboon, Goldhill and Starfish images.

5.4 Quality metrices

Segmentation quality of cuckoo search algorithm is obtained by quality metrices such as computational time, PSNR (peak signal to noise ratio) and SSIM (structural similarity index).

5.4.1 Computational time

Computational time specification can meet the demand of real-time image processing and **Tables 4** shows that the optimal output is reached with fewer iterations indicating less time to converge. Low run time is reached with Kapur fitness function winning in 13 out of 20 cases and otsu with 7 out of 20 cases. In general, Computational time increases as the level of threshold increases. But the CSA based MLT achieves the computation within a reasonable amount of time. Quick global convergence is achieved in less time by maintaining the diversity of population effectively. Levy flight with infinite mean and variance allows larger changes in magnitude and direction from the current position in search space.

Table 4. CPU time (s) of each algorithm

Images	No. of thresholds	Kapur method CSA	Otsu's method CSA	MCE method CSA
Lena	4	1.007780	0.977818	1.087521
	5	1.302429	1.342012	1.411131
	6	1.759438	1.706068	1.831942
	7	2.108483	2.205990	2.356025
Airplane	4	1.004558	1.044710	1.080848
	5	1.336676	1.374064	1.434904
	6	1.825396	1.717111	1.865030
	7	2.136269	2.409146	2.367477
Baboon	4	1.094111	0.968136	1.093525
	5	1.419862	1.335000	1.42456
	6	1.790348	1.713826	1.869365
	7	2.156000	2.229007	2.428548
Goldhill	4	1.076432	1.096310	1.163808
	5	1.489951	1.404934	1.479675
	6	1.864444	1.791522	1.963528
	7	2.272580	2.487100	2.380089
Starfish	4	1.024416	1.041476	1.054550
	5	1.341694	1.369150	1.542192
	6	1.750967	1.714934	1.857201
	7	2.170626	2.233899	2.324789

5.4.2 Power Signal to Noise Ratio

Peak Signal to Noise Ratio (PSNR) indicates visual depiction of segmented image from original image. CSA based MLT with high PSNR and low MSE confirms the accuracy.
PSNR is stated as

$$PSNR = 20\log_{10}\left(\frac{255}{RMSE}\right) \tag{31}$$

Where MSE refers the root mean-squared error,

$$MSE = \sqrt{\frac{1}{MN}\Sigma_{i=1}^{M}\Sigma_{j=1}^{N}\left[I(i,j) - I^{\wedge}(i,j)^{2}\right]} \tag{32}$$

where I and I^ refers the original and thresholded images and M×N indicates the dimensions of an image. Tables 5 show the comparison of CSA based MLT at 4, 5, 6 and 7 threshold level. High PSNR is obtained by Otsu objective function winning in 11 out of 20 cases, MCE with 6 out of 20 cases and 3 with Kapur out of 20 cases. Thus, from the Table 5, Otsu fitness function takes the upper hand over MCE and Kapur fitness function, indicating the object of interest to be inferred from the segmented image even with increase in number of thresholds.

Table 5. MSE and PSNR of each algorithm

Test Images	No. of thresholds	CS (Kapur method)		CS (Otsu method)		CS (MCE method)	
		MSE	PSNR	MSE	PSNR	MSE	PSNR
Lena	4	591.392	20.412	532.648	20.866	525.7723	20.9228
	5	578.014	20.511	418.394	21.915	502.2048	21.1220
	6	313.915	23.163	244.515	24.248	460.2345	21.9772
	7	391.56	22.203	215.319	24.800	242.5572	24.2827
Airplane	4	623.896	20.180	741.027	19.433	748.8105	19.3871
	5	1028.797	18.098	558.632	20.660	385.6384	22.2690
	6	675.242	19.836	405.885	22.047	166.811	23.9086
	7	389.084	22.230	236.71	24.389	282.0332	23.6278
Baboon	4	811.148	19.040	664.58	19.905	551.424	20.7159
	5	551.659	20.714	513.755	21.023	488.4427	21.2427
	6	392.047	22.197	375.535	22.384	366.6681	22.4881
	7	315.484	23.141	317.571	23.112	332.9727	22.9067
Goldhill	4	773.213	19.248	712.423	19.603	905.0834	18.5639
	5	998.845	18.136	580.024	20.496	953.5507	18.3374
	6	461.428	21.490	538.911	20.816	510.1676	21.0537
	7	392.175	22.196	334.509	22.887	476.0198	21.3546
Starfish	4	613.831	20.250	504.203	21.105	1080	17.8001
	5	515.907	21.005	392.097	22.197	799.4007	19.1032
	6	528.99	20.896	320.609	23.071	509.5304	21.0591
	7	462.917	21.476	269.534	23.825	314.6937	23.1519

5.5 Structural Similarity Index (SSIM)

Structural Similarity (SSIM) index measures the consistency between the true and segmented image. Higher value of SSIM by Otsu objective function winning in 15 out of 20 cases, Kapur with 3 out of 20 cases and 2 with MCE out of 20 cases, confirms the quality of original image. Thus, the Tables 6 infers the superior performance of CSA based Otsu with maximum value of SSIM compared to Kapur and MCE.

$$SSIM(x,y) = \frac{(2\mu_{\tau}\mu_{s} + c_1)(2\sigma_{\tau s} + c_2)}{(\mu_{\tau}^2 + \mu_{s}^2 + c_1)(\sigma_{\tau}^2 + \sigma_{s}^2 + c_2)}$$

μ_τ and μ_S are mean intensity of true and segmented image,
σ_τ and σ_S are the standard deviation of true and segmented image
$\sigma_{\tau S}$ indicates covariance of true and segmented image
c_1, c_2 are constants.

Table 6. SSIM of each algorithm

Images	No. of thresholds	CSA (Kapur method)	CSA (Otsu method)	CSA (MCE method)
Lena	4	0.9191	0.912	0.909
	5	0.9061	0.933	0.905
	6	0.8727	0.964	0.858
	7	0.9530	0.962	0.942
Airplane	4	0.7258	0.606	0.724
	5	0.7520	0.713	0.765
	6	0.7621	0.961	0.847
	7	0.7877	0.723	0.770
Baboon	4	0.8828	0.903	0.907
	5	0.8011	0.911	0.905
	6	0.9042	0.923	0.923
	7	0.8190	0.946	0.918
Goldhill	4	0.7575	0.767	0.684
	5	0.6577	0.733	0.646
	6	0.7787	0.915	0.778
	7	0.8003	0.813	0.774
Starfish	4	0.8762	0.904	0.809
	5	0.9056	0.908	0.880
	6	0.8968	0.918	0.903
	7	0.8963	0.926	0.926

6. Conclusion

In this paper, cuckoo search algorithm based MLT is used to attain the desired objective by overcoming the drawbacks of conventional color image segmentation such as premature convergence, inefficacy to find the points in near vicinity and getting stuck with suboptimal point. The simple, most widely used non-parametric objective functions such as Otsu, Kapur and MCE help to seek the optimal threshold to infer the details from the input image. CSA based MLT is tested with 5 standard test images at 4,5,6 and 7 threshold levels. Performance is authenticated through metrices namely objective values, optimal thresholds, CPU time, PSNR and SSIM.

The comparative analysis with experimental reults of MLT based CSA proves that the Otsu based cuckoo search algorithm outperformed than Kapur and MCE. Otsu based CSA achieves best in class performance in terms of CPU time, PSNR and SSIM. The intelligent approach of the controlling parameter 'Pa' with 75% search time for global optimal point and 25% search for local optimal point helps to attain our goal without any delay. The superiority of the proposed technique is that it employs Levy flight's infinite mean and variance to accomplish intensified and diversified search for wide range of applications.

References

[1] Liang H, Jia H, Xing Z, Ma J and Peng X 2019 Modified grasshopper algorithm based multilevel thresholding for color image segmentation, *IEEE Access,***7**, p 11258 –95.

[2] Jia H, Peng X, Song W, Lang C, Xing Z and Sun K 2019 Hybrid multiverse optimisation algorithm with gravitational search algorithm for multithreshold color image segmentation, *IEEE Access*,**7**, p 44903 –27.

[3] Cheng H D, Jiang X H, Sun Y and Wang J 2001 Color image segmentation: advances and prospects, *Pattern Recognition* **34**, p 2259-81.

[4] Upadhyay P and Chhabra J K 2019 Kapur's entropy based optimal multilevel image segmentation using Crow Search Algorithm, *Applied Soft Computing*, p 105522.
[5] Sathya P D and Kayalvizhi R 2011 Optimal multilevel thresholding using bacterial foraging algorithm, *Expert Systems with Applications*, **38(12)**, p.15549–64.
[6] Li J, Tang W, Wang J and Zhang X 2019 A multilevel color image thresholding scheme based on minimum cross entropy and alternating direction method of multipliers, *optik*, **183**, p.30-37.
[7] Yin P Y 2007 Multilevel minimum cross entropy threshold selection based on particle swarm optimization, *Applied Mathematics and Computation* **184(2)**, p 503–13.
[8] Sarkar S, Das S and Chaudhuri S 2014 A multilevel color image thresholding scheme based on minimum cross entropy and differential evolution, *Pattern recognition letters*. 54, p 27-35.
[9] Chen H, Jiang W, Li C and Li R 2013 A heuristic feature selection approach for text categorization by using chaos optimization and genetic algorithm, *Mathematical problems in engineering*, p 1-6.
[10] Manikantan KV and Yaradoni DKS 2012 Optimal Multilevel Thresholds based on Tsallis Entropy Method using Golden Ratio Particle Swarm Optimization for Improved Image Segmentation, *Procedia Engineering*, **30,** p 364–71.
[11] Jiang Y, Tsai P, Yeh W C and Cao L 2016 A Honeybee mating Based Algorithm for Multilevel Image segmentation using Bayesian theorem, *Applied Soft computing*, 52, p1181-90.
[12] Li J, Tang W and Zhang X 2018 Multilevel Thresholding Selection based on variational mode decomposition for Image segmentation, *Signal Processing*, 147, p 80-91.
[13] Baskan O 2013 Determining optimal link capacity expansion in road networks using cuckoo search algorithm with Levy flights, *Journal of applied mathematics*, p 1-11.
[14] Pare S, Kumar A, Bajaj V and G.K Singh 2016 A multilevel color image segmentation technique based on cuckoo search algorithm and energy curve, *Applied soft computing*, **47** p 76-102.
[15] Zhao X, Turk M, Li W, Lien K C and Wang G 2016 A multilevel thresholding segmentation algorithm based on two-dimensional K-L divergence and modified particle swarm optimization, *Applied soft computing*, **48** p 151-59.
[16] Bhandari A K, and Kumar I V 2019 A context sensitive energy thresholding-based 3D Otsu function for image segmentation using human learning optimization, *Applied Soft Computing*, **82**, p 105570.
[17] Pare S, Kumar A, Bajaj V and Singh G K 2017 An efficient method for Multilevel Colour image thresholding using cuckoo search algorithm based on minimum cross entropy *Applied Soft Computing* **61**, p 570-92.
[18] Yang X S and Deb S 2009 Cuckoo search via Lévy flights. World Congress on Nature & Biologically Inspired Computing (NaBIC 2009). *IEEE Publications* p 210–14.
[19] Tein L H and Ramli R 2009 Recent Advancements of Nurse Scheduling Models and A Potential Path Proceedings of the 6thIMT-GT Conference on Mathematics, Statistics and its Applications (ICMSA2010) Universiti Tunku Abdul Rahman, Kuala Lumpur, Malaysia.
[20] Layeb A 2011 A novel quantum inspired cuckoo search for knapsack problems. International Journal of Bio-Inspired Computation, **3(5)**, 297.
[21] Ouaarab A, Ahiod B and Yang X S 2013 Discrete cuckoo search algorithm for the travelling salesman problem, Neural Computing and applications, p 1659-69.

Performance comparison of IEEE 802.15.4 and IEEE 802.15.4e based MAC algorithm in wireless body sensor networks

Durga Prasad[1], Niranjan N. Chiplunkar[2] and K. Prabhakar Nayak[3]

[1]Department Electronics and Communication, NMAM Institute of Technology, Nitte, Karnataka, India
[2]Department Computer scinece and Engineering, NMAM Institute of Technology, Nitte, Karnataka, India
[3]Department Electronics and Communication, Manipal Institute of Technology, Manipal

[1]Email: durgaprasad@nitte.edu.in

Abstract. One of the potential applications of Wireless Sensor Network for improving the lifestyle of human beings is Wireless Body Sensor Network (WBSN). A dedicated health monitoring system is referred to as the Ubiquitous Health Monitoring (UHM) system consisting of the Internet of Things (IoT) based access points which are made available in a cloud-based server. So, ZigBee based IEEE 802.15.4 technology is extensively used in the implementation of Medium Access Control (MAC) and Physical (PHY) layers. Hence, a comparison is provided in terms of the features of the modified IEEE 802.15.4e standard. An algorithmic approach for the MAC layer helped in regulated power consumption of the sensor node by controlling the transmission power for the packets. This is initiated in the algorithm by categorizing the packets as Delay sensitive and Delay insensitive. By regulating the channel access strategy using Medium Access Control (MAC) layer, the energy consumption of the transceiver reduces, thus improving the energy consumption of the communication system and lifetime of the sensor network. For better throughput using IEEE 802.15.4e, the performance of Deterministic and Synchronous Multi-channel Extension (DSME) is found to be trustworthy than Time Slotted Channel Hopping (TSCH).

Keywords: MAC, PHY, Energy efficiency, throughput, DSME and TSCH.

1. Introduction

Wireless Body Sensor Network (WBSN) is formed by tiny body sensor nodes placed on specific locations on the human body which forward the precious physiological data into an access point called Body Area Network (BAN) coordinator. While performing the sensing, processing and communication operation, each node spends certain amount of energy [1]. Ubiquitous Health Monitoring (UHM) system consists of one or more WBSN system with body sensor nodes placed at specific areas on the body, measuring distinct parameters, forwarding the information to an access point. This system has an emerging technology with recent advancements in ICT which enables increased energy efficiency in the medical healthcare system as well as accuracy in the availability of data at multiple points. Figure 1 shows the block diagram of UHM System. There are WBSN systems in each room with an IoT based access point.

Each access point links to a cloud based system where vital data is stored. Thus, the global needs of healthcare system are satisfied with the state-of-the-art technology such as 'IoT' based access points and 'cloud storage' services [2].

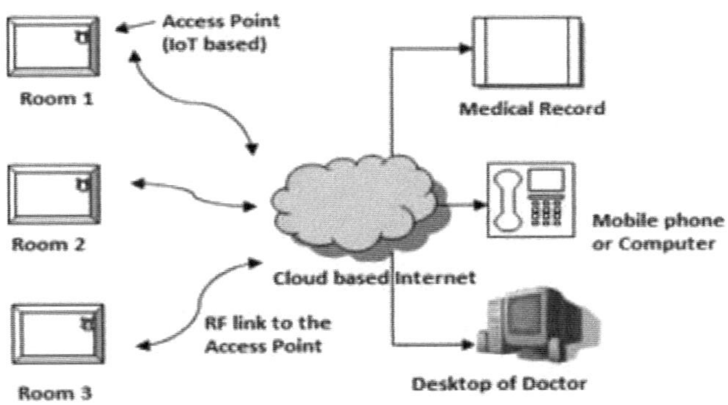

Fig 1. WBSN system as UHM system

The information can be downloaded to a medical record system, mobile phone or computer of a medical practitioner [3]. This procedure facilitates the necessary immediate action to be initiated by any of the people attached to these systems as each room might have a minimum of one patient attached with the Body sensor nodes. As the sensor nodes are small in size, there must be minimum power consumption during those operations so as to conserve the energy for longer period. The protocols of Network, Medium Access Control (MAC) and Physical (PHY) layers set up the rules and tasks in order to manage the operation of the WBSN system. The amplitude of the voltage or current pulses representing the (data bits) signal, are subjected to the power consumption of the communication unit of the PHY layer. The radio (transceiver) of the body sensor takes up the responsibility of data transmission and reception.

When the sensed data in the form of bits (PHY layer) are to be transmitted through Radio Frequency (RF) link to the destination node, initially channel access mechanism needs to be functional. The algorithm of MAC layer senses the channel and determines the channel status to be 'busy' or 'available'. Only if the channel status is 'available', the transmission will begin [18]. The radio of the PHY layer follows the instructions of MAC layer and switches the radio. If more than one-hop communication is required, then the decision of the network layer algorithm is needed as well. Hence, energy efficiency enhancement is contributed by the lower layers. The energy consumption of the radio chip must be well under the control of MAC and routing algorithm for increasing the life span of the sensor [4]. Depending on the regulation of power, the body sensor will have increased lifespan from months to year [5[. By solving few of the issues of the network management through algorithms for Medium Access Control (MAC) layer, energy efficiency improvement feature will be attended. An algorithmic approach for MAC layer proposed an increase in the life span of the Body sensor, and hence, network lifetime.

2. IEEE 802.15.4 based MAC algorithm

The Medium Access Control layer algorithms control the functions this layer. The physical layer performance is well managed by these algorithms and as such intended to provide high energy efficiency. The following standards of Wireless networks are more widely used in WBSN applications.

2.1. IEEE 802.15.4 standard

In this standard, the MAC and PHY layers are defined so as to meet the requirement such as high data rate, low power consumption, low latency, and high throughput. For Wireless Body Sensor Network, it uses the popular multiple access protocol of Wireless networks known as Carrier Sense Multiple Access with Collision Avoidance (CSMA/CA) [6, 7].

The following are some of the features of CSMA/CA used in WBAN.

a. Representation of data in super frame
b. Sophisticated data transfer model
c. Enhanced successful data delivery rate
d. Low power consumption leading to improvement in network lifetime
e. Security aspects in the protocols

The following sections describe a superframe structure of CSMA frame. A superframe has beacon period and control message fields as shown in Figure 2. Beacon is a short duration pulse used for marking the beginning of the frame. It is used for synchronization of the frame.

During the Contention Access Period (CAP), body sensors contend for channel access. This is referred to as channel access using CSMA/CA protocol.

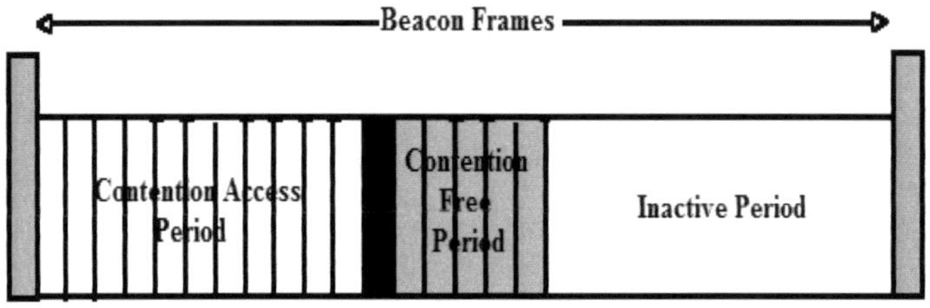

Fig 2. Superframe structure

In Contention Free Period (CFP), superframe structure has a dedicated time slot called Guaranteed Time Slot (GTS) for the transmission of data using TDMA protocol. In most of the emergency situations, this access mechanism is used for the transfer of data within low delay profile. In Inactive Period, the devices enter into low power (sleep) mode as energy saving strategy. Battery powered devices are operated using duty cycling to reduce power consumption. Each device periodically listens to the channel to check whether the message is pending or not. During this period, the sensor enters into "sleep" state whereas the radio is turned to standby mode.

The signal from PHY layer will be transmitted as bit through wireless channel. During the propagation, there will be fading of the signal which is estimated with fade depth distribution [8]. Latency is a prime parameter that decides the QoS of a WBSN system. The details of application packets received and the reason for failure of packets are provided. Greater the transmitting power, larger will be the signal to noise ratio and the range of transmission of the information. But this will lead to an increase in the power consumption of the node, reducing the efficiency and also life span of the body sensor node. Hence, a stipulated amount of transmitting power for the given BAN scenario should be estimated.

2.2 IEEE 802.15.4e Standard

In order to gratify the modern needs of industrial and healthcare sectors, IEEE 802.15.4 standard is added with an amendment to MAC layer and resulted as IEEE 802.15.4e standard. The new paradigm of IEEE 802.15.4e standard provides MAC extensions to different features. Few of them include, slotted access shared with dedicated slots, multi-channel communication, energy efficiency and frequency hopping. Most of the features are planned for supporting industrial sensor networks requiring internetworking services. There are two prime MAC behavioral modes utilising CSMA/CA or TDMA schemes. They are known as Time Slotted Channel Hopping (TSCH) and Deterministic and Synchronous Multi-channel Extension (DSME) respectively. A brief discussion of TSCH and DSME is given below [9, 19, 20].

(a) TSCH:

This MAC scheme is applicable for process automation scheme with a demand of focus on the equipment. It has time slotted access feature which is enhanced from the IEEE 802.15.4 providing a higher throughput and low latency. With a multichannel feature, larger communication with more sensor nodes is possible. This adds improvement to communication reliability and results in increased network capability. It also has channel hopping scheme with a benefit of interference mitigation. It is most suitable for multi-hop networks allowing star, mesh, multi-hop or tree topologies.

(b) DSME:

It supports industrial and healthcare applications requiring low latency and high flexibility. One of the key features of DSME is that it enables the MAC to operate in fully scheduled mode utilizing guaranteed time slots (GTS), termed as DSME-GTSs, for communications. The specialty of IEEE 802.15.4e superframe is the multi superframe structure with 3 or 4 superframes in it. Every one contains an Enhanced Beacon (EB), Contention Access Period (CAP) and Contention Free Period CFP. The CFP consists of slots that can be availed using GTS and contention based slots are available in CAP. Similar to IEEE 802.15.4 standard, EB will be transmitted at the beginning of slot -0 of the superframe. There are other features such as CAP reduction, group acknowledgement, distributed beacon scheduling and channel diversity modes.

Table 1 shows the comparison of MAC protocols of IEEE 802.15.4 and IEEE 802.15.4e standards. Under IEEE 802.15.4, as far as power consumption of radio is concerned, TDMA scheme provides better results than CSMA/CA at medium traffic levels. But, synchronization needs to be adjusted for every round of packet transmission. However, larger the distance between the sending and receiving nodes, larger will be the power consumption by the body sensor node.

Table 1: Comparison of MAC protocols of IEEE 802.15.4 and IEEE 802.15.4e

Performance parameters	IEEE 802.15.4		IEEE 802.15.4e	
	CSMA/ CA	**TDMA**	**TSCH**	**DSME**
Contentions	CAP or CFP	CFP	Fixed TDMA	CAP and Multiple CFP
Power Consumption	High	Low	Low	Depends on the number of nodes
Scalability	Medium	Poor	High	High
Traffic level	Low	Medium	Medium	High
Synchronisation	Not essential	Essential	Essential	Beacon enabled
Latency	Low	Medium	Bounded	Deterministic

Under IEEE 802.15.4e, as DSME provides multiple CFP slots, CAP reduction scheme, group acknowledgement and deterministic facility; it is preferred in WBSN applications over TSCH.

3. WBAN Specific Protocols

IEEE 802.15.4 using low rate WPAN standard protocol provides low energy consumption but low data rate. It used Slotted CSMA/CA- contention free MAC protocol. Node power consumption per packet increases at medium data packet arrival rate. But it introduces hidden node problem. Multi-hop topology or use of relays could be more suitable in such cases. Instead of transmitting the data packets from source to destination through direct transmission, it can be forwarded through neighbouring nodes so as to reach the destination node. This is referred to as Minimum Transmission Energy (MTE) routing scheme. However, the results have shown that the energy consumed by the MTE scheme is more than the direct communication method [10]. Instead, the formation of a small local network with the BAN head as

access point will be considered [11]. It is essential to note that, the management (control) packets that are helping to transmit the data packets are to be exchanged initially. This is an essential phase of the network operation as the body sensor nodes must aware of the resources, transmission and reception schedules, information on availability of channel and other relevant information. However, the control packet causes protocol overheads which enhance the duty cycle of the body sensor nodes. MAC operation must be adaptive to control packet management, access and link establishments to meet the requirements of duty cycle in body sensor devices operating in a particular band of frequency. For MICS band, a 36 seconds access to the channel in an hour is set [12].
It is required to provide an algorithmic approach to MAC layer that utilises the network topology based on the condition of the body parameters. It is essential to improve overall energy efficiency factor of WBSN system during the normal conditions as it is the most common case. The initial transmission of control packets must contribute to an improvised energy efficiency factor in this case. A design of MAC protocol for WBSN resulting in favour of energy efficiency for normal condition of the patient is desired. Initially, the algorithm is able to categorise the data into delay sensitive and delay insensitive packets. Delay insensitive packets are to be sent with lesser number of handshake signals from the buffer based on the priority of the packets. This transmits the normal and prioritised packets with lesser amount of energy spent by the transceiver. Finally, the algorithm is analysed with DSME scheme under IEEE 802.15.4e standard. The methods of energy calculation, algorithmic approach and performances of MAC schemes are discussed in the next sections.

4. Energy Efficient Algorithmic Approach for MAC Layer

Low-Energy Adaptive Clustering Hierarchy (LEACH) algorithm is a protocol under hierarchical routing category. It makes use of several energy reduction concepts that are necessary for energy conservation of WBAN system. A LEACH protocol utilizes TDMA schedule based MAC algorithm for communication between the nodes inside the cluster tree [13]. Also it considers minimised overhead, and hence, contributes for low power consumption. It involves communication with cluster heads and thus avoiding over emitting problem [14].

4.1. Cluster Topology

The solutions such as routing scheme, clustering, novel Cluster Head (CH) selection and data aggregation bring down the energy consumption of the sensor nodes. Figure 3 illustrates the WBSN, comprising CH and Cluster Member (CM). The CH collects the data from all CM and transmits to the sink. The clustering approach provides a solution for monitoring the large Region of Interest (RoI). The monitoring of small areas can be achieved by Layered architecture.

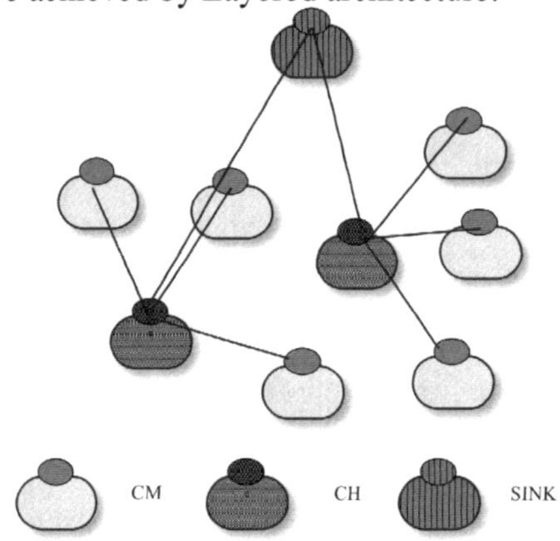

CM – Cluster Member CH- Cluster Head

Fig 3. WBSN architecture

The monitoring application classifies data packets into Delay Sensitive (DS) and Delay Insensitive (DIS). The so called DS applications are not always delay sensitive in nature. In this case, the number

IOP Conf. Series: Materials Science and Engineering 1119 (2021) 012020 doi:10.1088/1757-899X/1119/1/012020

of bits of handshake signal can be reduced when the network is monitoring DIS packets. This can limit the energy consumed by the network. Figure 4 illustrates the classification of packets in the event monitoring system.

4.2. ALGORITHM

```
Input
State {DS, DIS}
Begin
  While:
          If state = = DIS
                Store data till the buffer limit;
           If (buffer limit = = full)
           {
             Reduce the no. of bits of handshake signal;
     Start transmission;
    If buffer contains priority
packet
           Prioritize;
           else
             fifo;
       else state = = DS;
     Follow routing;
}
```

The above algorithm is simulated using IEEE 802.15.4 standard and later with IEEE 802.15.4e standard. The observations are presented in the next section.

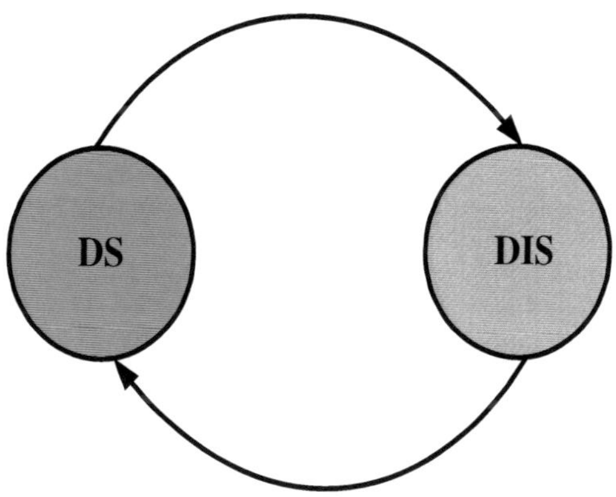

Fig 4. Classification of packets in event monitoring

The transition between DS and DIS monitoring is achieved through Markov model. Markov is a memoryless model in which the transition totally depends on the present state and not on the previous state [15].

5. Results

The proposed method is simulated in MATLAB with 9 nodes. The proposed algorithm is validated with LEACH protocol for DS packet transmission. For Delay insensitive packets, IEEE 802.15.4 and 15.4e standards are used. The handshake bits are varied with fixed buffer capacity and the difference of energy consumed is monitored. Table 2 illustrates the simulation prelims considered in evaluating the approach.

Table 2: Simulation prelims

Parameter	Value
Network size	500 × 500 m
Number of nodes	9
Initial energy of sink	2 J
Propagation model	Two ray ground
Packet size	4000 bits
Transmission energy	8.281x10^{-5}J/bit
Receiving energy	3.5x10^{-5} J/bit
Packet generation rate	0.02–0.2 kbps

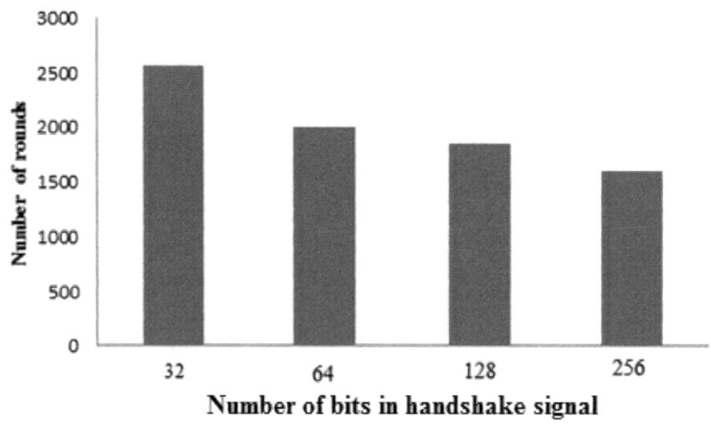

Fig 5. Network lifetime

Figure 5 illustrates the lifetime comparison with respect to handshake signal size. The proposed model provides high lifetime with low handshake signals. The data arrival rate is assumed to be following Poisson's distribution. As the number of bits of handshake signal is reduced, the proposed model provides good network lifetime.

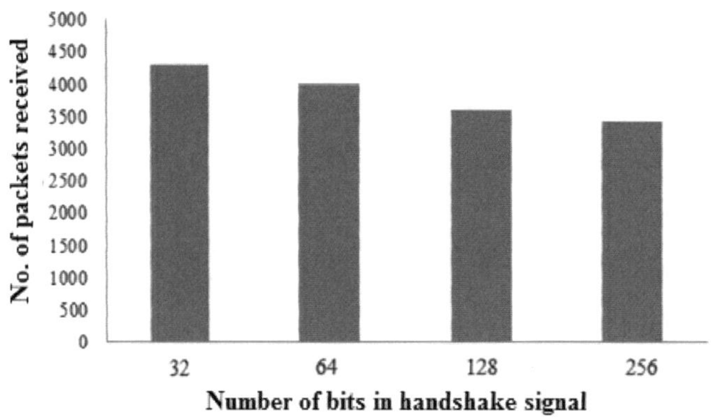

Fig 6. Network throughput

Figure 6 illustrates the network throughput of the proposed algorithm. From the Figure 6, it is clear that the proposed MAC layer approach transmits large number of packets with smaller handshake bits. IEEE 802.15.4e based DSME MAC provides extensive features for the applications of WBSN requiring internetworked access point and hence, suitable for related applications in future. Hence, the algorithm is tested analytically using the probabilistic models of [16] and [17]. The throughput is analysed for increased number of sensor nodes as given in Table 3.

Table 3: Comparison of Throughput for IEEE 802.15.4 and 802.15.4e standards

No. of sensor nodes	Normalised Throughput	
	IEEE 802.15.4	IEEE 802.15.4e
6	0.303	0.44
8	0.342	0.52
10	0.358	0.553

Table 3 compares the throughput of the proposed algorithm using IEEE 802.15.4 and IEEE 802.15.4e standards with the number of sensor nodes. It is clear from the comparison that, in IEEE 802.15.4e based DSME MAC mode, a higher throughput is achieved than with IEEE 802.15.4 standard.

Table 4: Comparison of Energy consumption for IEEE 802.15.4 and 802.15.4e standards

No. of sensor nodes	Energy consumption	
	IEEE 802.15.4(J)	IEEE 802.15.4e(J)
6	1.427×10^{-4}	9.44×10^{-5}
8	1.450×10^{-4}	9.47×10^{-5}
10	1.469×10^{-4}	9.52×10^{-4}

Table 4. compares the Energy consumption of the proposed algorithm using IEEE 802.15.4 and IEEE 802.15.4e standards with the number of sensor nodes. It is clear from the comparison that, the transceiver of sensor node consumes lower power with IEEE 802.15.4e based DSME MAC mode, than with IEEE 802.15.4 MAC standard. Hence, IEEE 802.15.4e standard provides enhanced energy efficiency factor than its earlier version. Hence, the algorithm provides an enhanced throughput for IEEE 802.15.4e standard than its earlier version.

6. Conclusion

Wireless Body Area Network is formed by establishing a duplex RF link between the Body sensor nodes and a BAN coordinator or access point. From the literature survey, it is studied that an amendment made on the existing IEEE 802.15.4 standard gives rise to IEEE 802.15.4e standard. Comparison of the two MAC modes namely TSCH and DSME shows DSME as the most preferable mode for WBSN. Features of DSME such as enhanced energy conservation strategies with improved reliability and latency attracts the MAC designers of WBSN. In the MAC layer's algorithmic approach, the data packets are categorized into DS and DIS types. The DIS packets are transmitted with reduced number of bits based on the priority of data. By reducing the overhead bits or handshaking signal bits, the energy consumption of the radio has been reduced by 0.1 percent per cycle of data transfer. For DS type of packets, the cluster tree based algorithm with slotted TDMA scheme is incorporated. Hence, the proposed algorithm provides energy efficiency improvement based on the delay sensitive or delay insensitive category of packets. For WBSN, it is practically easier to implement the MAC modes available of IEEE 802.15.4e rather than with IEEE 802.15.4 standard. In future, a Ubiquitous healthcare system will be constructed with the sensor nodes desiring to get connected to internet based devices using IEEE 802.15.4e based MAC layer algorithm. Thus, the performance of DSME form of MAC algorithm in IEEE 802.15.4e proves to be the suitable technology in terms of throughput and life time enhancement of the WBSN system in place of IEEE 802.15.4 based MAC algorithm.

References

[1] Negra, R., Jemilia I., and Belghitha, A. Wireless Body Area Networks: Applications and Technologies. The 2nd International Workshop on Recent Advances on Machine-to-Machine Communications, Procedia Computer Science 83, 2016, 1274 – 1281.

[2] Hassanalieragh, M., Page, A., Soyata T., Sharma, G., Aktas, M., Mateos, G., Kantarci B., and Andreescu, S. Health Monitoring and Management Using Internet-of-Things (IoT) Sensing with Cloud-based Processing: Opportunities and Challenges. IEEE International Conference on Services Computing, 2015, 285 – 292.

[3] Shah, R. C., and Yarvis, M. Characteristics of On-body 802.15.4 Networks. In 2nd IEEE Workshop on Wireless Mesh Networks, (WiMesh), 2006, 138 – 139.

[4] Zaman, N., Tang Jung, L. and Yasin, M.M., 2016. Enhancing energy efficiency of wireless sensor network through the design of energy efficient routing protocol. Journal of Sensors, 2016.

[5] Hao, Y., and Foster, R. Wireless Body Sensor Networks for Health-Monitoring Applications in Physiological Measurement. 2008, Vol. 29, No. 11.

[6] Hiep, P. T., Hoang, N. H., and Kohno, R. Performance Analysis of Multiple-Hop Wireless Body Area Network. Journal of Communications and Networks, 2005, Vol. 17, No. 4, 419 – 427.

[7] Kaitalidou, D. S., Boulogeorgos, A. A., and Pavlidou, F. N. Comparison of CSMA/CA Protocols Applied in Wireless Body Area Network Standards. Proceedings of the 18th Mediterranean Electro Technical Conference (MELECON), Cyprus, 2016, 1 – 6.

[8] Tse, D. and Viswanath, P., 2005. Fundamentals of wireless communication. Cambridge university press.

[9] De Guglielmo, D., Seghetti, A., Anastasi, G., & Conti, M. (2014, June). A performance analysis of the network formation process in IEEE 802.15. 4e TSCH wireless sensor/actuator networks. In 2014 IEEE Symposium on Computers and Communications (ISCC) (pp. 1-6). IEEE.

[10] Heinzelman et al. (2000) Heinzelman, W. R., Chandrakasan, A. and Balakrishnan, H. Energy-Efficient Communication Protocol for Wireless Microsensor Networks. Proceedings of the 33rd Hawaii International Conference on System Sciences, 2000.

[11] Kanagachidambaresan & Chitra (2015) Kanagachidambaresan, G. R., and Chitra, A. Fail Safe Fault Tolerant Mechanism for Wireless Body Sensor Networks (WBSN). Wireless Personal Communication, Springer, 2015, 80, 247 – 260.

[12] Karuppiah et al. (2016) Karuppiah Ramachandran, V., Ayele, E., Meratnia, N. and Havinga, P., 2016. Potential of wake-up radio-based mac protocols for implantable body sensor networks (ibsn)—a survey. Sensors, 16(12), pp. 2012.

[13] Kaur & Mahajan (2012) Kaur, S., and Mahajan, L. Power Saving MAC Protocols for WSNs and Optimization of S-MAC Protocol. International Journal of Engineering & Business Management. Special Issue on Radio Frequency Identification and Wireless Sensor Networks, 2012, Vol. 4.

[14] Long (2013) Long, H., Routing and MAC based on LEACH and SMAC for Energy Efficiency and QoS in Wireless Sensor Networks. Doctoral thesis, University of Huddersfield, 2013.

[15] Baynat et al. (2019) Baynat, B., Bu, G. and Potop-Butucaru, M., 2019. Markovian model for Broadcast in Wireless Body Area Networks. arXiv preprint arXiv:1906.05103.

[16] Park, T.R., Kim, T.H., Choi, J.Y., Choi, S. and Kwon, W.H., 2005. Throughput and energy consumption analysis of IEEE 802.15. 4 slotted CSMA/CA. Electronics Letters, 41(18), pp.1017-1019.

[17] Sahoo et al.(2017) Sahoo, P., Pattanaik, S. and Wu, S.L., 2017. A reliable data transmission model for IEEE 802.15.4e enabled wireless sensor network under WI-FI interference. Sensors, 17(6), pp. 1320.

[18] Velusamy, Bhanumathi, and Sangeetha Chitteth Pushpan. "An enhanced channel access method to mitigate the effect of interference among body sensor networks for smart healthcare." IEEE Sensors Journal 19, no. 16 (2019): 7082-7088.

[19] Kurunathan, Harrison, Ricardo Severino, Anis Koubaa, and Eduardo Tovar. "IEEE 802.15. 4e in a nutshell: Survey and performance evaluation." IEEE Communications Surveys & Tutorials 20, no. 3 (2018): 1989-2010.

[20] Chiti, Francesco, Romano Fantacci, and Andrea Tani. "Performance evaluation of an adaptive channel allocation technique for cognitive wireless sensor networks." IEEE Transactions on Vehicular Technology 66, no. 6 (2016): 5351-5363.

IOP Conf. Series: Materials Science and Engineering 1119 (2021) 012021 doi:10.1088/1757-899X/1119/1/012021

System Model of VANET using Optimization-Based Efficient Routing Algorithm

S Kannadhasan[1], M Shanmuganantham[2], R Nagarajan[3]
[1]Assistant Professor, Department of Electronics and Communication Engineering, Cheran College of Engineering, Tamilnadu, India
[2]Lecturer (S.G), Department of Electrical and Electronics Engineering, Tamilnadu Government Polytechnic College, India
[3]Professor, Department of Electrical and Electronics Engineering, Gnanamani College of Technology, Tamilnadu, India
[1]kannadhasan.ece@gmail.com

Abstract: - A wireless networking strategy is the VANET (Vehicular ad hoc network). This approach is used to connect between vehicles and to raise the question of traffic quality. The most important aspect is the connection between V2v and V2i, so it is essential to ensure appropriate protection and privacy techniques. VANET safer, effective and avoid illegal vehicle interference with other cars, this work suggests an authentication scheme based on the hash feature. Only an authenticated vehicle may use the party key agreement scheme using the Chinese residual theorem. The forward and backward confidentiality was done in this manner. Security and privacy are fulfilled by our scheme. The most favorable aim is to create more useful, productive and safer roads across vehicle networks by educating the basic authorities and drivers in the future on a timely basis. Another aim is to identify wireless technology for the development of VANET. The aim is to protect and render feasible commercial requests by means of a variety of short to medium communication systems and/or other VANET networks. This technology will help the primary concern for safe contact in crucial time and meet the QOS criteria of other interactive multimedia applications or e-commerce.
Keywords: - VANET, System Model, Established Route and Node Pair

1. Introduction

VANET seeks to offer appealing facilities such as protection services, including alerts of curve level, notifications of emergency vehicles, lane shift assistance, warnings of pedestrian walking, warnings of traffic breaks, warnings of road crossings, and warnings of road condition. In addition, weather details, traffic information, the position of the gas station or restaurant and other conveniences and interactive facilities such as Internet connectivity may also be offered. The Intelligent Transportation System (ITS), the growing amount of automobiles in the region, is one of the most promising recommendations for the effective management of urban transport [1]-[5]. The VANETs were introduced by researchers and vehicle manufacturers on the basis of the Mobile Ad-hoc Network (MANET) concept to develop next-generation transport systems. VANET is a distributed, less infrastructural, self-organizing communication network that is developed between moving vehicles as a variant of MANET.

VANET implies Ad hoc vehicle network which is the infrastructure used to transfer vehicles and build a transportable network as a joint throughout the network. Participating vehicles become a VANET wireless link or router to enable the vehicles to attach almost 100 to 300 meters to each other and other vehicles are linked to each other in order to establish a broad range network such that the mobile internet is created. For protection purposes, it is believed that the first networks to adopt this technology are fire and police mobiles to communicate with each other [6]-[8].

In VANET or Intelligent Vehicular Ad-Hoc Networking, genius methods of utilizing vehicular networking are described. Multiple VANET-integrated ad-hoc networking technologies such as ZigBee, IRA, WiMAX IEEE, and Wi-Fi IEEE for easy, effective, reliable, quick and simple connectivity inside active mobility vehicles. Useful procedures such as media contact inside vehicles may be permitted, as well as the process of trailing automotive cars is often preferred. In cars, protection measures are characterized by VANET, flowing automotive contact, edutainment and telemetry. Wireless infrastructure collection is expected to be introduced as DSRC (Dedicated Short Range Communication) in VANET, which is used in Wi-Fi. Satellite, WiMAX, and Cellular are additional incoming wireless innovations [9]-[12].

Vehicular Ad-hoc Networks (VANET) may be viewed as ITSS equipment (Intelligent Transportation Systems). Vehicular networks were conceived by the Intelligent Transportation Systems. IVC (Inter-Vehicle Communication) enables cars to connect with each other at the same time as RVC (Roadside-to-Vehicle Communication) makes way side stations. The benefit of proactive routing protocols is that because the destination route is processed in the past, no route exploration is needed, but the drawback of this protocol is that it allows low latency for real-time application, it often contributes to the management of unused data routes, allowing the usable bandwidth to be limited. TORA is part of the link reversal routing family in which a cyclic graph is designed that guides the packet flow and guarantees its capacity to enter all nodes. The directed graph will be built by a node by broadcasting question packets. If the node has a downward connection to the destination when receiving a question packet, it will broadcast a return packet; otherwise, it will simply drop the packet. A reply packet receiving node can only change its height if the height of the answered packet is the minimum of all other reply packets [13]-[15]. The drawback of the TORA Algorithm is that it offers a path to network nodes, but it is impossible for VANET to manage all these routes. In intra-zone communication, a constructive routing protocol (IARP) is used in ZRP, and in intra-zone communication, an inner-zone reactive routing protocol (IARP) is used. If all are in the same routing field, the source sends data directly to the target, otherwise the IERP reactively initiates a path discovery. There are two levels of routing in HARP: intra-zone and inter-zone, based on the destination location. In intra region and inter-zone routing, respectively, it utilizes constructive and reactive protocols. It is not acceptable to ad hoc networks of high versatility.

2. System Model of VANET

The suggested algorithm discovers a perfect route to a source and sends the message using the route that it has identified. To figure out the optimal path, it utilizes the multi-objective bio-inspired heuristic cuckoo search optimization algorithm. The steps as defined for finding the best path. The VANET model is used, based on the following assumption, to choose the best routes between the source and its destinations. Network routing is a particular category of network that offers a number of resources to its users, such as message notifications, driver forecasts, and information services.

The road network has played a very important role in many ITS-related applications, such as brake warning, weather forecasts, traffic and automobile protection. There is also a significant necessity for a vehicle-to-vehicle contact device that is secure, stable and entitled to include a safety warning in due course. There have been many efforts to standardize the design and deployment of road networks. The network model is to be considered for the road model layout to be built. Let N be the statistic for the average number of network vehicles containing a number of segments. The route exploration approach here offers the origin with awareness of the destination direction.

3. Optimization-Based Efficient Routing Algorithm

For VANET routing, the easiest way to pick a path is typically to forecast traffic between the source and destination locations. For efficient route selection, the choice of a low traffic density is preferred as a high traffic density improves the movement of vehicles on the lane. Therefore, an active prediction approach is important for choosing the right route. The key step in the current system of application is path discovery. To provide the details effectively, the suggested algorithm selects the best multipath routing. It is important to configure the parameters for the purpose of successful route detection. The nest size and mutation probability must be determined to initialize the parameters for the upper and lower bands. It is important to select a suitable nest

$$\beta=\beta_{max}-(N_{iter}/N_{Iter(Total)})*(\beta_{max}-\beta_{min})\text{-----------}[1]$$

where, β_{max} and β_{min} represents the maximum and minimum nest size

N_{iter} represents the current iteration number

$N_{Iter(total)}$ represents the total iteration number respectively.

As expressed in the equation-1 nest size will decreases with increase in the iteration number. According to MOBHCSNO-ER algorithm mutation probability is associated with the fitness,

$$P_f=\{P_{fmin}+(P_{fmax}-P_{fmin})\}\text{-----------}[2]$$

where f=fitness (I) - $f_{min,}$

Which depends on the current quality of Ith solutions; fitness of the population, respectively l; P_{fmax} and P_{fmin} represent the maximum and minimum of mutation probability p_a, respectively. From the equation-2 it can be seen that the fitness of the solution gas adjusted and the mutation probability can generally vary with respect to the amount of the iteration. After the parameter gets initialized the nest position gets analyzed.

$$\alpha_v=\left(\gamma\left(1+\beta_p\right)\right)*\sin\left(\pi*\beta_p/2\right)/\left(\gamma\left(1+\beta_p 2\right)*\beta_p{}^{\beta p-1/2}\right)*\left(\beta p-1/2\right)^{1/(\beta p-1/2}\text{)-----------}[3]$$

where $\sigma_{v.}$ represents the random size of the nest.

The equation-3 can be rewritten as,

$$n_p=\text{rand}\left(\tau_{san},1\right)*\left(u_b-l_b\right)+l_b\text{ ----------}[4]$$

when n_p represents the random position of the nest

After that the objective function needs to be computed to find out nearest node path. The objective of node localization is to estimate the coordinates of the unknown nodes based on the anchor vehicular nodes. Separable each unknown node and its anchor node can be calculated.

$$\sigma^2=\gamma^2*e_{ij}^2\text{ ---------}[5]$$

Where σ^2 is the error variance and e_{ij}^1 is the original distance of the unknown node.

The measured distance between the unknown distance node and the anchor node was represented by using the equation-6

$$e_{ij'}=e_{ij}=N_{ij}\text{ ---------}[6]$$

IOP Conf. Series: Materials Science and Engineering 1119 (2021) 012021 doi:10.1088/1757-899X/1119/1/012021

where N_{ij} represents the error o the unknown vehicular node.

Then the objective function should be calculated which is a combined mean error of the unknown node and the anchor nodes.

$$f\left(x_i\ y_i\right)=1/n \sum_{j=1}^{n} (e^{ij}-e^{ij'})\hat{}2\text{----------}[7]$$

The unknown vehicular node can be estimated by its co-ordinates by running the MOBHCSNO-ER algorithm. The alpha, beta and gamma locations are then changed. It is easy to locate the path after upgrading it. Multiple routes would be sent to the source after the route answer is received from the endpoint. Each node's remaining capacity, load and hop count are attached to the reply packet. The node of origin then analyses the response packets received from multiple routes and calculates the level of fitness for each path direction. The highest degree of exercise and the n route positions is that pick the number of paths n. A source node routing table that has the fitness values of each of the routes in the descending order of the data transmitted through the fitness route. If the route succeeds, the next fitness merit path transmits it. Then, if all n paths are struggling to relay knowledge, then according to the algorithm, it begins to locate and repeat the process. If REPLY sends via the destination node to the neighbor node before the send hits that node, the routing table is changed by dragging it up until the receiver vehicle node is reached. The fitness value at the node of the sender vehicle is now calculated by applying the formula described in the algorithm.

The downward order of the routing table and the data is to be processed in the downward order type. In reality, the sender vehicle node now starts sending data according to the route that has the highest fitness value in the sender vehicle node routing chart, if the path fails, if the sender sends it with the second best fitness path, etc. Where F is the fitness value of every route that the source node has obtained and is to graded. The residual energy for each of the nodes and the distance of the bandwidth between each of the nodes can be determined here. In order to evaluate health, the power, metric delay and shortest route are used. Below is the approach to the technique proposed? The data packet sent by the source vehicular node here is the cuckoo's egg. Here, along with the data packets, the vehicular source is sent via the multi objective route that can be transmitted to the designator node. Data is passed by inadequate power or fallen out of heavy traffic routes.

Input: Vehicle Node-Vn, Node Coordinates-Vc, Maximum Bound-Vx, Vy, Energy Parameter-Ve.
Output: Priced-φp and optimized-γii (best nest and route path)
Step 1: initiates the parameters,
    $[V_x, V_y]$= [Upper Limit lower limit]
    Max - iteration max $_{I\,ter}$=100;
    Lower band $l_b=V_x(1)*(V_{c,}, 2)$;
    Upper band u_b=max (V_c, 2);
    Best near n_b Dimensions d_b=size (u_b, 2);
Step 2: initialize the nest position n_b,
    N= size (n_p)
    Beta _ pos β_p=3/2;
    $\sigma_v = (\gamma(1=\beta_p))*\sin(\pi*\beta_p/2)/(\gamma(1+\beta_p/2)*\beta_p{}^{\beta}{}_p{}^{-1/2})*(\beta_p=1/2)^{1/(\beta_p{}^{-1/2})}$
    for j=1: n
    s= n_p (j, :)
    u= randn (size(s)*σ_v)
    v= rand (size(s))
    Let compute nest position, n_p=rand (τ_{san}, 1)*(u_b-l_b) + l_b

4. Results and Discussion

Create a TA (Trusted Authority) to track the vehicle approaching the relevant area of coverage and provide the customer with authentication. If the source vehicle has no path to the destination vehicle, the route exploration is initiated in an on-demand manner by the source vehicle. The node looks up its own neighbour table after creating RREQ to find out whether it has a neighbour vehicle closer to the destination vehicle in Figure 1.

The RREQ packet is redirected to that vehicle if a closer neighbour vehicle is available. If the RREQ packet is not a closer neighbour car, all adjacent vehicles are flooded. Only in the following three situations does a destination vehicle respond to a received RREQ packet with a Path Reply (RREP) packet:

1. If the RREQ is the first packet to be obtained from this source car,
2. If the RREQ packet includes a higher source sequence number than the RREQ packet to which the destination vehicle has previously replied,
3. If the RREQ packet includes the same source sequence number as the RREQ packet that the destination vehicle has previously referred to, however the new packet suggests that a higher quality route is usable, is shown in Figure 2

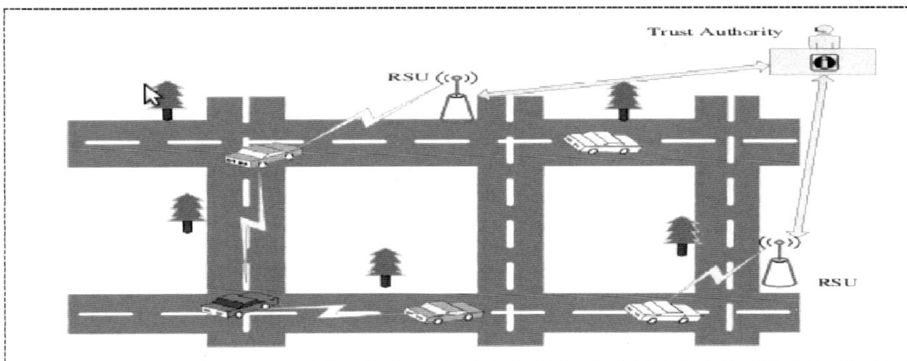

Figure 1. Roadmap of VANET System Model

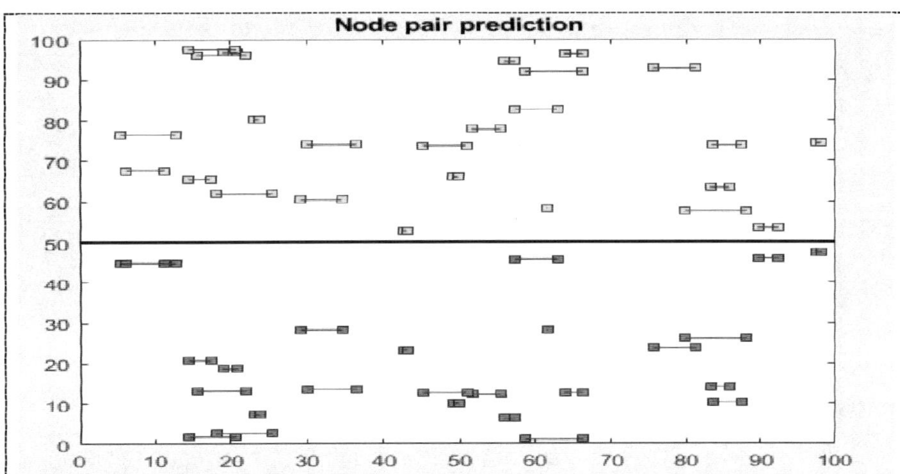

Figure 2. Node Pair Prediction

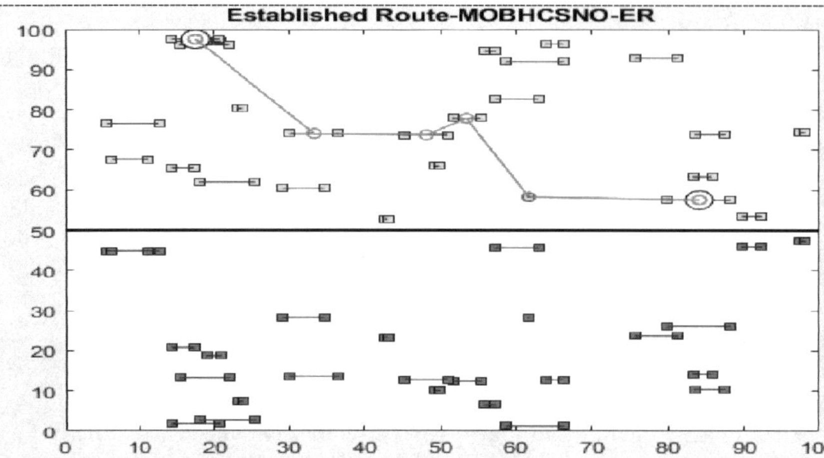

Figure 3. Established Route MOBHCSNO

Both VANET users should have their data reported with the RSU. The RSU can provide the customer with one initial packet key after registration. The consumer can get details about the other surrounding vehicles from the TA using this initial packet key. We're adding a routing protocol in this module to pass messages via RSU'S between the vehicles. Such correspondence should be service-oriented in order to take account of the different forms of data from the RSU. The RSU is a system for wireless networking that uses the DSRC protocol. It is a way of coordination between the car and the TA. Vehicles in the same RSU range also allowed confidential knowledge to be shared. In Figure 3, the contact between the vehicle and the facilities is primarily between the vehicle and an RSU fixed on the side of the lane. The short-range wireless communication protocol, called the dedicated short-range communication protocol, regulates all vehicle to vehicle and vehicle to infrastructure communications, which increases the overall protection and reliability of the transport system. Through unreliable wireless networking media, RSU links to TA. RSU is not fully trusted; TA needs to control and track it. Vehicle to vehicle connectivity is seen in Figure 4 only by the RSU. Without support from the RSU, automobiles do not interact explicitly with each other.

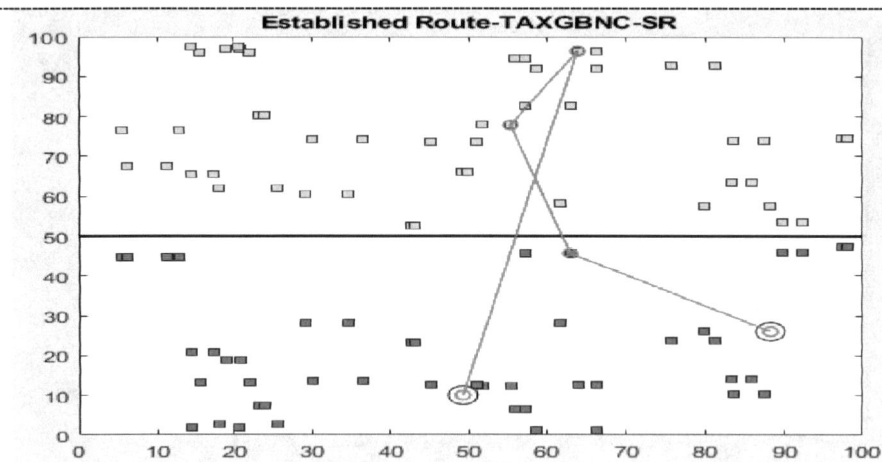

Figure 4. Established Route TAXGBNC

5. Conclusion

In urban settings, routing inside VANETs is a dynamic activity. Because of the enhanced mobility in the nodes, routing difficulties will arise. Therefore, the reasons of the reduced efficiency of the facilities would find the problem impossible to minimize. To substitute this problem with the usage of the Stochastic Discriminate Random Forest Node Classifier and TAXGBNC-SR, if the solution is to weighted end-to-end delay is to be used. A shortest secure route has been calculated here using this methodology. Finally, the findings of the simulation revealed that the Stochastic Discriminate Random Forest Node Classifier can provide the radio, latency, throughput and execution period of the packet distribution that performs well, indicating the method's effectiveness.

References

1. Tomar, R.S., Sharma, M.S.P., Jha, S., &Chaurasia, B.K. (2019). Performance Analysis of Hidden Terminal Problem in VANET for safe Transportation system. In harmony Search and Nature Inspired Optimization Algorithm-Springer, Singapore, pp. 1199 -1208.
2. Loganathan , G. B.(2019). Vanet Based Secured Accident Prevention System. International Journal of Mechanical Engineering and Technology, 10(6).
3. Karimzadeh Motallebiazar, M., Mariano de Souza, A., Zhao, Z., Braun, T., Villas, Z., L., Sargento, S., & Loureire, A.A. (2019). Intelligent Safety Message Dissemination with Vehicle Trajectory Density Prediction in VANETs., IEEE Transactions on Vehicular Technology Special Issue on Vehicular Networks in the era of 6G: IEEE
4. Koti, R.B., & Kakkasageri, M.S. (2019, march). Intelligent Safety Information Dissemination Scheme for V2V Communication in VANETs. 2019 IEEE International Conference on System, Computation, Automation and Networking (ICSCAN), pp. 1-6.
5. Jose, a. a., Pramod, A., Philip, G., & George, S.J. (2019) .Sybil attack detection in vanet using spidermonkey technique and ecc. International Journal of Wireless Communications and Networking Technologies, 8(3).
6. Nguyen, V., Khanh, T.T., Oo, T. Z., Tran, N.H., Huh,E.N., & Hong, C.S. (2019). A Cooperative and Reliable RSU-Assisted IEEE 802.11 P-Based Multi Channel MAC Protocol for VANETs. IEEE Access, 7, pp.107576-107590. 10.1109/ACCESS.2019.2933241
7. Tomar, R.S., Sharma, M.S.P., Jha, S., &Sharma B. (2019). Vehicles Connectivity-Based Communication Systems for Road Transportation Safety, Soft Computing: Theories and Applications, pp 483-492, Springer, Singapore.
8. Li, W., Song,W., Lu, Q., & Yue, C. (2020). Reliable Congestion Control Mechanism for Safety Application in Urban VANETs, Ad Hoc Networks, 102033. https://doi.org/10.1016/j.adhoc.2019.102033
9. Naja, A., Boulmalf, M., & Essaaidi, M. (2019). Toward a New Broadcasting Protocol to Disseminate Safety Messages in VANET Environment, International Conference on Mobile, Secure, and Programmable Networking, pp. 163-172. Springer, Cham.
10. Benkirane, S. (2019, April). Road Safety against Sybil Attacks Based on RSU Collaboration in VANET Environment. International Conference on Mobile, Secure, and Programmable Networking, pp. 163-172. Springer, Cham.
11. Cui, J., Wu, Zhang, J., Xu, Y., & Zhong, H. (2019). An Efficient Authentication Scheme Based on Semi-Trusted Authority in VANETs. IEEE Transaction on Vehicular Technology, 68(3), pp. 2972-2986.
12. S.Kannadhasan, G.Karthikeyan and V.Sethupathi, A Graph Theory Based Energy Efficient Clustering Techniques in Wireless Sensor Networks. Information and Communication Technologies Organized by Noorul Islam University (ICT 2013) Nagercoil on 11-12 April 2013, Published for Conference Proceedings by IEEE Explore Digital Library 978-1-4673-5758-6/13 @2013 IEEE.

13. Erskine, S. K., & Elleithy, K. M. (2019). Real Time Detection of DoS Attacks in IEEE 802.11 p Using Fog Commuting for a Secure Intelligent vehicular network, Electronics, 8(7), 776 MDPI.
14. Limbasiya, T., & Das, D. (2019). ESCBV: energy-efficient and secure communication using batch verification scheme for vehicle users, Wireless Networks, 25(7), pp. 4403-4414, https://doi.org/10.1007/s11276-019-02104-5
15. Hasrouny, H., Samhat, A. E., Bassil, C., & Laouiti, A. (2019, April). A Trusted Group-Based Revocation Process for Intelligent Transportation System., International Conference on Digital Economy, pp. 133-146. Springer, Cham.

IOP Conf. Series: Materials Science and Engineering 1119 (2021) 012022 doi:10.1088/1757-899X/1119/1/012022

Mid-infrared supercontinuum generation using low peak pump power in $As_{38}Se_{62}$ based chalcogenide photonic crystal fiber

S Vyas[1(a)], N S Shiong[2], A Kumar[1], G Sharma[1], Jaiverdhan[1], V K Chandna[1], G Singh[3]

[1] Electronics and Communication Engineering Department, Jaipur Engineering College & Research Centre (JECRC), Jaipur, 302022, India
[2] Institute of Nano Optoelectronics Research and Technology (INOR), Universiti Sains Malaysia, Penang, 11800, Malaysia
[3] Electronics and Communication Engineering Department, Malaviya National Institute of Technology (MNIT), Jaipur, 302017, India

[(a)] sandypvyas@gmail.com

Abstract. A supercontinuum (SC) based chalcogenide $As_{38}Se_{62}$ PCF for broadband mid-infrared light source is numerically reported. For the computational studies, the design of the proposed structure is made up of three rings of air holes with circular and elliptical shapes. The proposed structure provides excellent nonlinear coefficient and dispersion optimization. For the analysis, the finite difference frequency domain (FDFD) method is employed. Due to the high nonlinear refractive index and optimizing design of $As_{38}Se_{62}$ chalcogenide glass, an effective mode area of 40.5972 μm^2 is obtained. The dispersion characteristic of the proposed structure has a zero-dispersion wavelength at 3.89 μm. The nonlinear coefficient is 761 $W^{-1}km^{-1}$ at the wavelength of 4 μm. Dispersion is almost flat from 2 μm up to 10 μm. The supercontinuum spectrum calculated ranges from 2 μm to 9 μm. The presented structure is appropriate for medical imaging, optical coherence tomography and optical communications.

1. Introduction

Researchers have studied holey optical fibers or photonic crystal fibers (PCFs) for nearly two decades. The established optical fiber is made up of a cladding and a core. In traditional optical fiber systems, the cladding refractive index is taken lower as compared to the core refractive index. Light passes within the center of the fiber due to the total internal reflection. In the last few years, there has been rapid development in the field of fiber devices and photonic crystal fibers. Photonic crystal fibers are mainly of two types, one hollow-core PCF and the other solid-core PCF. The PCF is formed by an unconventional cladding structure, with periodic patterns of air-vents throughout the fiber. The effective refractive index is reduced by air vents and the coupled light is directed within the core [1-3].

Due to the unique properties of PCFs, there are many applications in the field of sensors and fiber light sources compared to conventional fibers or bulk media. Changes in the size, shape, and position of vents in the microstructured cladding as well as the appropriate choice of a foundation material permit the transmission, dispersion and even nonlinear properties of fibers to be controlled. The dispersion characteristic of any PCF depends on its internal structure. The main feature of PCF is its unique dispersion behaviour, which includes ultra-flat dispersion characteristics, zero-dispersion wavelength shifting, as well as chromatic dispersion optimization [4-5].

Higher-order dispersions including third-order dispersion along with several nonlinear processes, namely, self-phase modulation, self-steepening, cross-phase modulation, stimulated Brillouin scattering, four-wave mixing and stimulated Raman scattering are responsible for supercontinuum (SC) generation. It is typically evolved by bringing out ultra-short laser pulses (picoseconds, femtoseconds) in extremely nonlinear holey optical fibers with injecting close to zero-dispersion wavelength (ZDW). The transfer of ZDW is done using the modification in geometric parameters of the PCF appropriately, allowing supercontinuum to be generated in a desideratum spectral belt. In the modern-days, SC light source is being used in many scopes namely telecommunication, nonlinear microscopy, optical metrology, cosmological studies, ultrashort pulse generation and optical coherence tomography. The PCF is the most appropriate means for producing an effective SC due to ZDW and group velocity dispersion (GVD) can be controlled in PCF. Dispersion is a major factor for the generation of the SC spectrum, with the help of this creating better SC generation in a certain division. The first condition to achieve a broad spectrum is a low and flat dispersion curve. This can be acquired in PCF geometry by making appropriate changes in the circumference of the core, the circumference of the air vents and the gap between air vents [6-7].

Due to the utility of supercontinuum light sources, a lot of progress has been made in this area over the years. However, SC technology is still in its early days and requires more work beyond 5 micron IR frequencies for industrial applications. Highly nonlinear soft glasses are used for better results in the mid-IR division, namely chalcogenide (As_2Se_3, As_2S_3, GeAsSe), heavy metal oxide, Tellurite (TeO_2), and ZBLAN. Chalcogenide glass-founded PCFs (made of S, Se, or T) have high-quality photonic utility compared to all other available infrared fibers for nonlinear applications. The broad transmission window, controlled dispersion in infrared range, and large nonlinearity up to hundred times higher than ZBLAN and silica glasses make chalcogenide glass the most suitable material for SCG [8-9].

Non-silica glasses are used for applications in the mid-infrared range, where As_2Se_3 and $As_{38}Se_{62}$ chalcogenide glasses are more prevalent because their nonlinear coefficient (~100 times larger) is higher than silica. $As_{38}Se_{62}$ glass attracts researchers more because of its distinct properties, it has higher stability against crystallization during drawing than As_2Se_3. SC spectrum broadening cannot achieve more than 2.5 μm in silica material due to absorption, so it is not suitable for the mid-IR region. $As_{38}Se_{62}$ has a transparency of 2 to 10 μm, in addition, it is a low loss material and has a glass transition temperature (Tg) of 165° C which is suitable for SCG in the mid-IR region [10-11].

Several studies have reported that $As_{38}Se_{62}$ chalcogenide glass is a low-loss material and is suitable for SCG. Mazhorova et al. demonstrated $As_{38}Se_{62}$ glass capillaries used as waveguides in the mid-IR range, and fabrication of capillaries by drawing techniques. These capillaries showed low-loss transmission in mid-IR light [10]. Venck et al. fabricated $As_{38}Se_{62}$ glass microstructured optical fiber by casting method and obtained ZDW at 4.84 micron. In this research, the glass has an absorption peak of about 4.56 micron because of the appearance of Se–H chemical bonds [12]. Numerical and experimental effects on SCG have been presented by Møller et al. in the $As_{38}Se_{62}$ material from 1.7 to 7.5 μm with 320 fs pulses having ZDW at 3.5 μm [13]. Recently, Ghosh et al. reported on SC expanding from 3.1 to 6.02 micron and from 3.33 to 5.78 micron by pumping at 4 micron and 4.53 micron, respectively [8].

The presented article shows the desired result by modifying the dimension of the air hole in the cladding as required. Here a solid-core PCF is designed to have a dispersion graph in the anomalous zone, with the help of which a chalcogenide-based material produces a broad SC in the mid-infrared zone. The article is organized in the following order. Section 2 describes linear, nonlinear parameter modelling and SC generation. Section 3 introduces the PCF structure fabricated with $As_{38}Se_{62}$ chalcogenide glass. This section discusses numerical simulations and dispersion plots, nonlinear multiplier and effective mode area. In the next part 4, supercontinuum generation with designed PCF is presented. The investigation of SC generation dealing with the generalized nonlinear Schrödinger equation is also studied in this section. The last segment establishes the results.

2. Numerical model of generation of supercontinuum in PCF

The generalized nonlinear schrödinger equation (GNLSE) is used to study SC generation, taking into account the loss, stimulated Raman scattering, high order dispersion and frequency dependence of nonlinear operate. GNLSE is demonstrated as [5]

$$\frac{\partial A}{\partial z} + \frac{\alpha}{2}A - \left(\sum_{n\geq 2}\beta_n \frac{i^{n+1}}{n!}\frac{\partial^n A}{\partial t^n}\right) = i\gamma\left(1 + \frac{i}{\omega_0}\frac{\partial}{\partial t}\right)A(z,t)\int_{-\infty}^{\infty} R(t')|A(z,t-t')|^2 dt' \quad (1)$$

where A (z,t) represents electric field envelope, calculated using the split-step Fourier method, α represents linear loss, time is represented by t, ω_0 represents carrier frequency, β_n represents n^{th} derivative of the propagation constant β, z represents distance, τ represents retarded time travelling at the envelope group velocity and γ represents the nonlinear coefficient, expressed as [14]

$$\gamma = \frac{n_2\omega_0}{c\,A_{eff}} \quad (2)$$

where c denotes speed of the light in the vacuum, n^2 represents nonlinear refractive index of fiber $As_{38}Se_{62}$ is 1.13×10^{-17} ($m^2.W^{-1}$) [4]. The expression of A_{eff} denotes the effective cross sectional of the fiber to the wavelength as [15]:

$$A_{eff}(\lambda) = \frac{\left(\iint_{-\infty}^{\infty}|E|^2 dx\,dy\right)^2}{\iint_{-\infty}^{\infty}|E|^4\,dx\,dy} \quad (3)$$

where E is electric field distribution of the optical fundamental mode of $As_{38}Se_{62}$ PCF cross section. The delayed Raman response function R(t) in given equation can be written as [16]

$$R(t) = (1-f_r)\delta(t) + f_r\frac{\tau_1^2+\tau_2^2}{\tau_1\tau_2^2}\exp\left(-\frac{t}{\tau_2}\right)\sin\left(\frac{t}{\tau_1}\right) \quad (4)$$

The equation includes the Raman contributions of instantaneous electronic and vibrations $\delta(t)$. Parameter τ_1 relate the inverse of the phonon oscillation frequency and the parameter relates τ_2 is the bandwidth of the Raman gain spectrum of the material. The values of the parameters applied in this work for $As_{38}Se_{62}$ chalcogenide glass: τ_1 = 23.14 f_s, τ_2 = 157 f_s and f_r = 0.1, represent the fractional contribution of the delayed Raman Response [8].

3. Design of the $As_{38}Se_{62}$ chalcogenide glass PCF

In this section, we have worked on the design of the $As_{38}Se_{62}$ PCF for the study of optical properties. The main objective behind designing PCF is to achieve high nonlinearity and small and flat dispersion in the anomalous dispersion zone. The figure (1) exhibits a cross-section view of the introduced fiber. The cladding area is made up of three rings of different sized air holes. The first ring and the second ring are formed by circular air vents of different diameters so that better light confinement can be obtained in the central area. Elliptical-shaped air vents have been applied to achieve low and flat dispersion in the third ring. The parameters of air vents in the cladding are tabulated in Table 1. The difference between two adjacent holes is determined by the pitch.

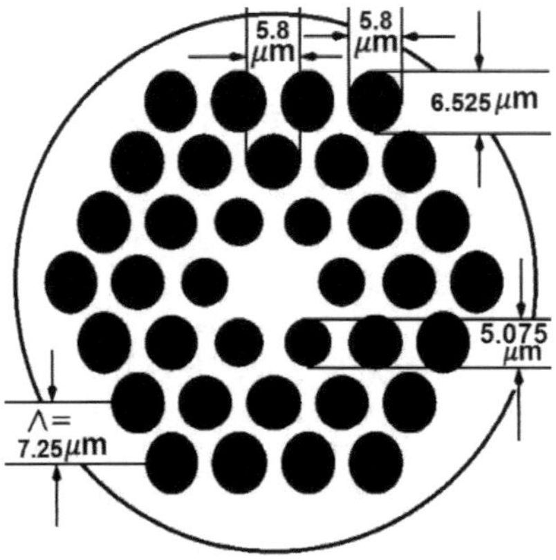

Figure 1. The introduced $As_{38}Se_{62}$ PCF structure

Table 1. Design parameters of fiber introduced.

Ring	Diameters Air- Holes	Pitch (Λ)
Ring-1	5.075 μm	
Ring-2	5.8 μm	7.25 μm
Ring-3	d_y= 6.525 μm d_x= 5.8 μm	

The geometric parameters of the PCF have been chosen such that with their help the dispersion in the anomalous zone and the wavelength of zero-dispersion near the inject wavelength available in the mid-infrared zone can be achieved. The Sellmeier equation is employed to determine the refractive index as a function of wavelength, using the following Sellmeier equation for $As_{38}Se_{62}$ [8].

$$n^2(\lambda) = 3.7464 + \frac{3.9057\lambda^2}{\lambda^2-(0.4073)^2} + \frac{0.9466\lambda^2}{\lambda^2-(40.082)^2} \quad (5)$$

The finite-difference time-domain (FDTD) mechanism has been applied to investigate the optical effects of the designed PCF. The sum of the current waveguide dispersion D_W (λ) and material dispersion D_M (λ) in any fiber determines the chromatic dispersion D_C (λ) of the fiber [5].

$$D_C(\lambda) = D_M(\lambda) + D_W(\lambda) \quad (6)$$

$$D(\lambda) = -\frac{\lambda}{c}\frac{d^2}{d\lambda^2}\mathrm{Re}\,[n_{eff}(\lambda)] = -\frac{2\pi c}{\lambda^2}\beta_2(\lambda) \quad (7)$$

where λ represents the operating wavelength, n_{eff} represents effective refractive index of the fiber and β represents propagation constant.

Numerical simulations of the optimized PCF design have achieved a broad dispersion over the anomalous zone as portrayed in figure (2). A wavelength of zero-dispersion at 3.89 μm has been obtained with dispersion in the anomalous region that is suitable for pumping the PCF with a peak pulse at the inject wavelength of 4 μm.

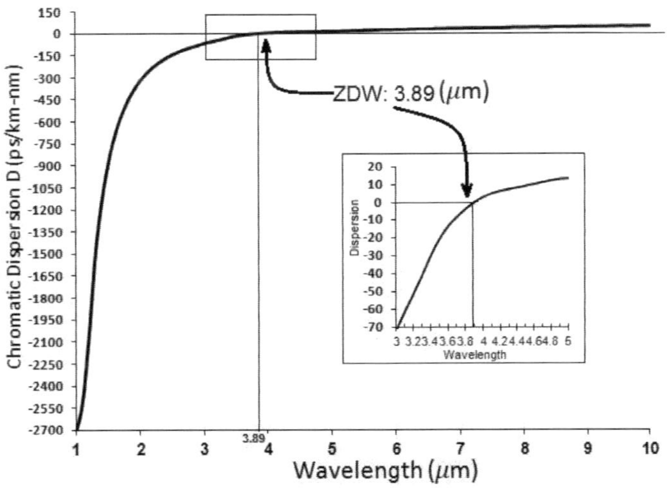

Figure 2. Dispersion curvature for fiber with ZDW at 3.89 μm

The two important parameters for obtaining a comprehensive supercontinuum are the nonlinear multiplier and the effective area obtained from the fundamental mode of fiber. This part simulated a numerically optimized design to obtain nonlinear coefficients and effective area values and is plotted in figure (3) for wavelengths taking the obtained results. At a inject wavelength of 4 μm, the nonlinear multiplier is evaluated as $(\gamma) = 761\ W^{-1}\ km^{-1}$ and the effective mode area $(A_{eff}) = 40.5972\ \mu m^2$.

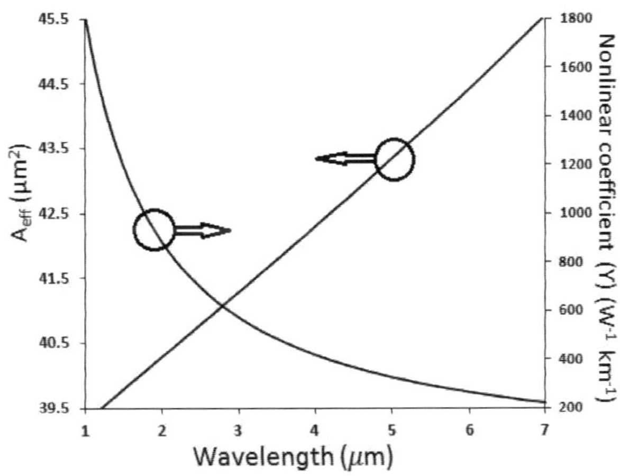

Figure 3. The nonlinear multiplier and effective mode area in the $As_{38}Se_{62}$ fiber

4. Simulations of SC in proposed fibers

The generation of supercontinuum is a complicated phenomenon that involves the creation of a specific combination of multiple nonlinear effects occurring under high-intensity pumping, in consequence broadband spectral coverage, spatial coherence and high-power spectral density. In this segment, we will discuss the generation of supercontinuum. After calculating both the non-linear and linear parameters of the introduced fiber at a inject wavelength of 4 μm, including chromatic dispersion in the anomalous dispersion region close to the wavelength of zero-dispersion. The extrication of GNLSE has been obtained by the symmetric split-step Fourier method for input pulses in both the time and frequency domains. The electric field distribution of the propagating mode is demonstrated by figure (4). Figure (5) depicts spectral evolution in fiber length of 1.33 kW, and 252 f_s input pulse having a secant hyperbolic pulse profile shown as [5,11]

$$A(z=0,t)=\sqrt{P_0 \operatorname{sech}\frac{t}{t_0}}\exp\left(-i\frac{c}{2}\frac{t^2}{t_0{}^2}\right) \quad (8)$$

where P_0 represents peak power and T_0 represents input soliton time span. The wide-band SC spanning from 2000 nm to 9000 nm was obtained by pumping the inject at 4 μm pump wavelength was obtained from 200 mm long PCF. The spectral growth of the 1.33 kW input pulse for various distance is displayed in diagram (6).

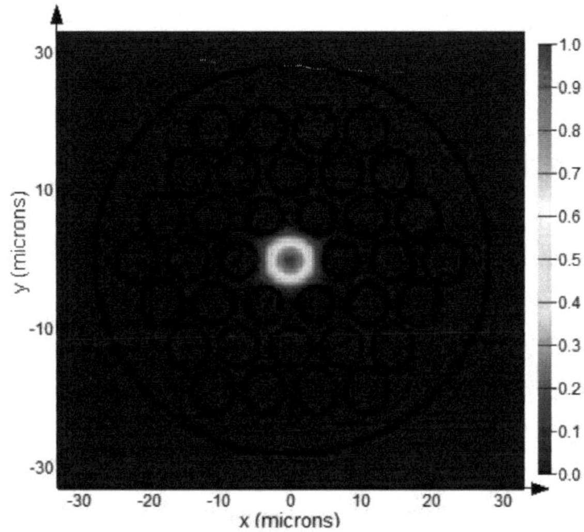

Figure 4. Electric field distributions in $As_{38}Se_{62}$ PCF

Some nonlinear processes are mostly responsible for the generation of supercontinuum in anomalous dispersion zone namely self-phase modulation, stimulated Raman scattering, soliton dynamics, and dispersive wave. The process of forming supercontinuum is mainly in two phases, in the first phase spreading because of self-phase modulation (SPM) and the second phase expanding due to stimulated Raman scattering, four-wave mixing and dispersion. After a spread of short distance (mm), the spectrum is split over the pair of peaks, after exiting the initial pump wavelength the number of sub-peaks between the pair of peaks continues to increase, resulting in the generation of spectra [16].

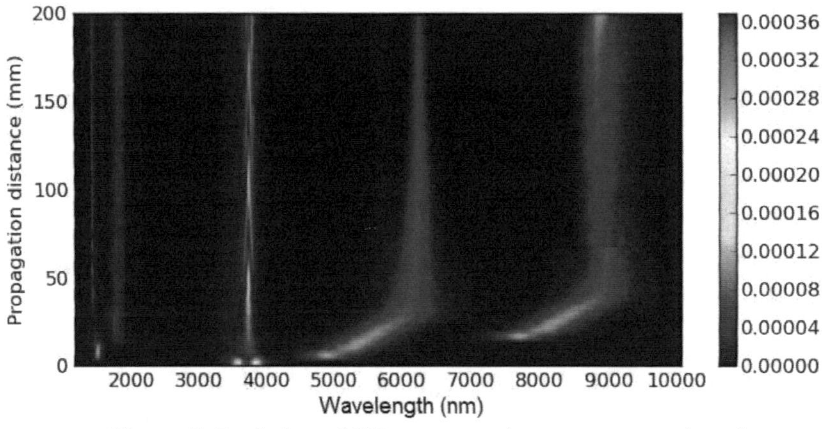

Figure 5. Evolution of SC spectra at 4 μm pump wavelength

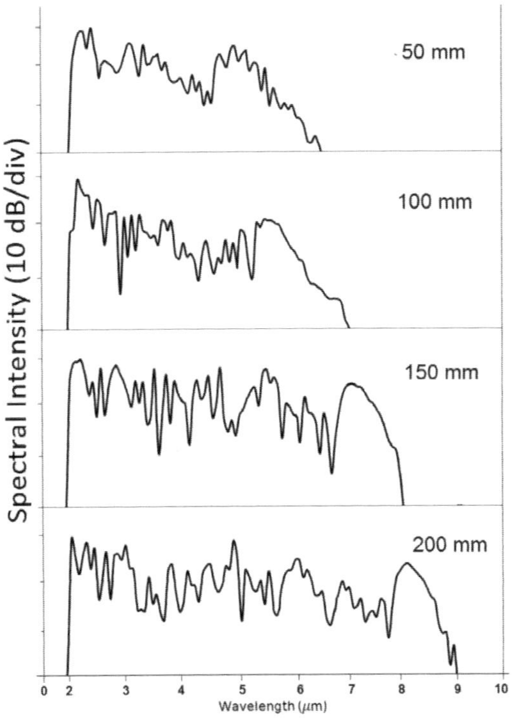

Figure 6. Spectral widening received in $As_{38}Se_{62}$ fiber with various distances

5. Conclusion

The proposed PCF design has resulted in high nonlinearity of 761 $W^{-1}Km^{-1}$ with a significantly effective area of 40.5972 μm^2. Extensive SC spectra can be generated even with the use of low pump power from high nonlinearity. Results from simulations of $As_{38}Se_{62}$ chalcogenide photonic crystal fibers with a length of 200 mm to SC generation 2 to 9 μm have been achieved. To achieve this, a low pulse peak power of 1.33 kW has been used while taking 252 f_s as the pulse duration. The expansion of the SC spectrum depends on the dispersion profile of the fiber and its input characteristics, which have been achieved by changing the PCF parameters. The use of this obtained broadband SC is in the fields of optical metrology, telecommunications, cosmological studies, nonlinear microscopy, optical coherence tomography, and ultra short pulse generation.

References

[1] Sharma M, Konar S 2016 *Optics Communications* **380** 310-319
[2] Pathak P, Zafar R, Kanungo V, Vyas S 2020 *Journal of Optical Communications* **1**
[3] Vyas S, Tanabe T, Singh G and Tiwari M 2016 *2016 International Conference on Computational Techniques in Information and Communication Technologies (ICCTICT)* (India: Delhi/IEEE) p 607-611
[4] Diouf M, Salem A B, Cherif R, Wague A, Zghal M 2016 *Optical Materials* **55** 10-16
[5] Vyas S, Tanabe T, Singh G and Tiwari M 2016 *Ukrainian journal of physical optics* **17** 132-139
[6] Dinh Q H, Pniewski J, Van H L, Ramaniuk A, Long V C, Borzycki K, Xuan K D, Klimczak M, and Ski R B 2018 *Applied Optics* **57** 1-9
[7] Gupta V, Kanungo V, Kalia P, and Vyas S 2020 *International Conference on Innovative Advancement in Engineering and Technology (IAET-2020)* (India: Jaipur/SSRN) p 3548424
[8] Ghosh A N, Meneghetti M, Petersen C R, Bang O, Brilland L, Venck S, Troles J, Dudley J M, and Sylvestre T 2019 *J. Phys.: Photonics* **1** 044003

[9] Vyas S, Tiwari M, Tanabe T and Singh G 2016 *Proceedings of the International Conference on Recent Cognizance in Wireless Communication & Image Processing* (New Delhi/Springer,) p 409-413
[10] Mazhorova A, Markov A, Ung B, Rozé M, Gorgutsa S, and Skorobogatiy M 2012 *Journal of the Optical Society of America B* **29** 2116-2123
[11] Vyas S, Tanabe T, Tiwari M and Singh G 2016 *Chinese optics letters* **14** 123201
[12] Venck S, St-Hilaire F, Brilland L, Ghosh A N, Chahal R, Caillaud C, Meneghetti M, Troles J, Joulain F, Cozic S, Poulain S, Huss G, Rochette M, Dudley J and Sylvestre T 2020 *Laser & Photonics Review* **14** 1-20
[13] Møller U, Yu Y, Kubat I, Petersen CR, Gai X, Brilland L, Méchin D, Caillaud C, Troles J, Luther-Davies B and Bang O. 2015 *Optics express* **23** 3282-91
[14] Vyas S, Tanabe T, Tiwari M, Singh G 2016 *2016 International Conference on Advances in Computing, Communications and Informatics (ICACCI)* (India: Jaipur/IEEE) p 2521 – 2526
[15] Kalra S, Vyas S, Tiwari M, Buryy O and Singh G 2018 *Acta Physica Polonica A* **133** 1000-1002
[16] Kalra S, Vyas S, Tiwari M, Ismail Y, Yupapin P, Ali J and Singh G 2020 *Journal of Modern Optics* **67** 920-926

AUTHOR INDEX